Standard Grade
Physics

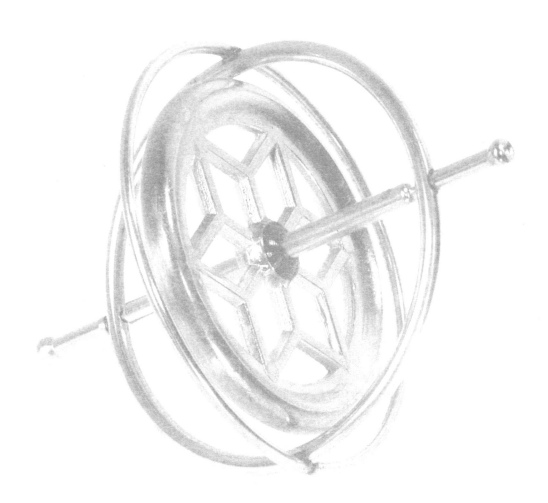

John Taylor ✕ Brian Arnold

Contents

Telecommunication

Using electricity

Health physics

Electronics

Transport

Energy matters

Space physics

Sound and light waves

Telecommunication

We live in an information age. Telecommunication is when information is sent (**transmitted**) and taken in (**received**) over a far distance.

Thunder and lightning

We see the lightning flash, then we hear the thunder. Light travels so fast, it takes virtually no time at all to reach us! The **speed of light** is almost **1 million times faster than the speed of sound**.

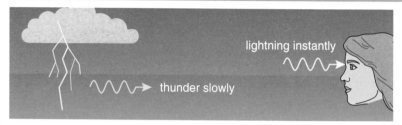

lightning instantly

thunder slowly

velocity of light, v_{light} = 300 000 000 m/s = 3×10^8 m/s.

velocity of sound, v_{sound} = 340 m/s = 3.4×10^2 m/s.

We can work out how far away the thunderstorm is using the speed of sound alone.

speed is defined as the **distance travelled in unit time** (1 s).

$$\text{Speed} = \frac{\text{Distance}}{\text{Time}} \qquad v = \frac{d}{t} \qquad \text{where } v = \text{speed in m/s}$$

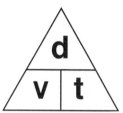

d = distance travelled in m t = time taken in s

e.g. During a thunderstorm there was a delay of 20 s between the lightning flash and the clap of thunder. How far away was the storm?

$d = v \times t = 340 \times 20 = 6800$ m.

Sound and light signals are essential in telecommunication.

Top Tip
Sound needs particles to travel through; light can travel in a vacuum.

Measuring the speed of sound

We can measure the speed of sound in air in the lab.

- Measure a distance, say 1 m.
- Place 2 microphones this distance apart attached to a **fast timer** or computer and interface.
- The timer starts timing when the sharp sound of the hammer passes microphone 1 and stops timing when the sound passes microphone 2.

hammer   bang

plate

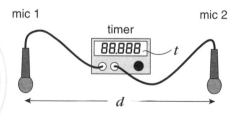

mic 1

mic 2

timer

- Then calculate using $v = \dfrac{d}{t}$.

Waves ... are a movement of energy

In telecommunications we see how we can communicate (**transmit signals**) a long distance using waves: **light waves / electric waves / radio waves / microwaves**.

Waves on the ripple tank

A ripple tank lets us see all the main properties of waves. The small motor makes the rod vibrate, the rod creates ripples and the lamp helps us see the wave pattern on the floor.

If the rod is made to vibrate more quickly, the frequency of the waves increases, but their wavelength decreases.

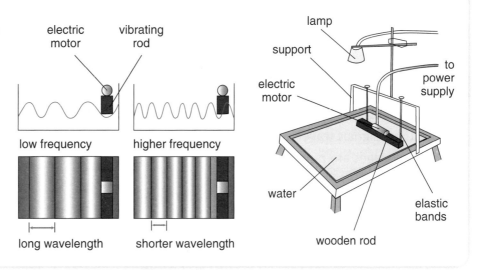

Measuring a wave

Wave property	Definition	Formula	Unit
Wavelength λ	The distance till a wave repeats itself.		Metres, m
Amplitude a	The height from the rest position.		Metres, m
Frequency f	The number of waves in unit time. $\text{Frequency} = \dfrac{\text{number}}{\text{time}} = \dfrac{1}{\text{period}}$	$f = \dfrac{N}{t}, f = \dfrac{1}{T}$	Hertz, Hz
Period T	The time for 1 wave (to pass).	$T = \dfrac{1}{f}$	Seconds, s
Velocity v	The distance travelled in unit time.	$v = \dfrac{d}{t}$	m/s

1 wavelength λ will pass in 1 period T of time, with a velocity $v = \dfrac{\lambda}{T}$ but $T = \dfrac{1}{f}$ therefore the **wave equation** gives the velocity as $v = f\lambda$ m/s.

This is a very useful equation to remember for waves as they are often going too fast to measure using $v = \dfrac{d}{t}$.

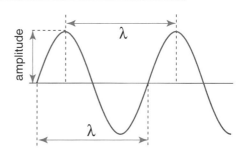

Quick Test

1. How long does sound take to travel 1 km?

2. How long does light take to travel 1 km?

3. 10 waves are 5 m in length. What is their wavelength?

4. From peak to trough a wave measures 60 cm. What is its amplitude?

5. A wave has a frequency of 10 Hz. What does this mean?

6. There are 20 waves passing a point in 4 s . What is the wave a) frequency, b) period?

7. A wave of frequency 5000 Hz and length 2 cm is travelling at what speed?

The telegraph to the telephone

Top Tip
The signals in electrical cable travel at almost the speed of light – almost 300 000 000 m/s

The Morse code telegraph

- Coded messages are sent out by a **transmitter** (tap key).
- Electrical signals travel along long electric wires.
- The messages are picked up by a **receiver** (buzzer or speaker).

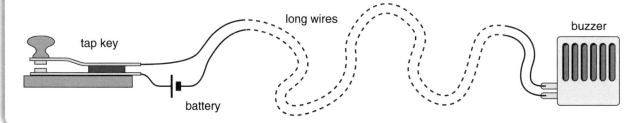

The telephone

Earpiece – Receiver – Loudspeaker

Electrical Energy ⟶ Sound Energy

    Mouthpiece – Transmitter – Microphone

    Sound Energy ⟶ Electrical Energy

During a telephone conversation **electrical signals are transmitted along wires** and these can be examined using a **Cathode Ray Oscilloscope** (called a **CRO** or **oscilloscope** for short).

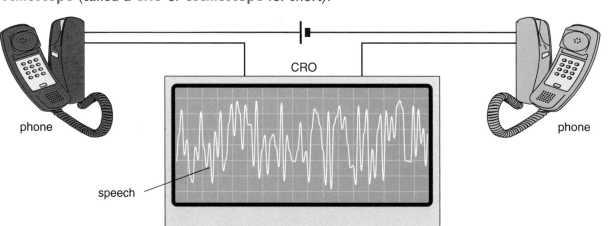

Irregular **waves** are seen on the screen. Speech is a mix of notes or waves.

The signals in electric cable travel at almost the speed of light – almost 300 000 000 m/s.

Signals on the oscilloscope

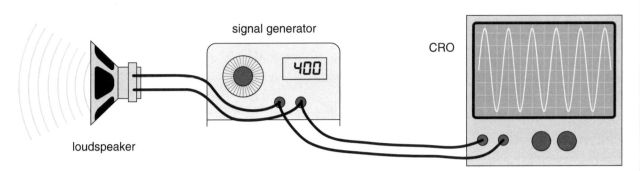

signal generator

CRO

loudspeaker

The **signal generator** creates pure signals.
The **loudspeaker** turns electrical signals to sound.
The **oscilloscope** examines the signals on a screen.

Adjust the **volume**:

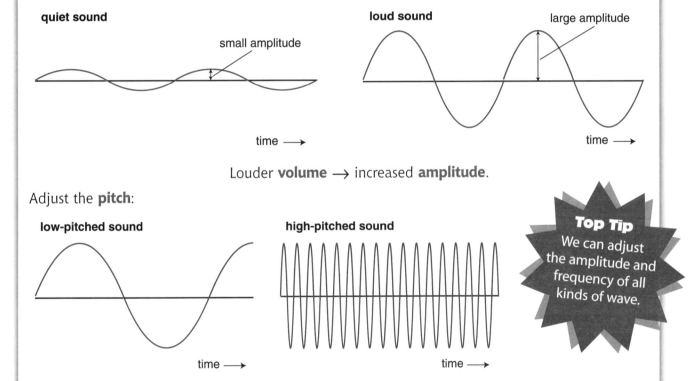

quiet sound

small amplitude

time →

loud sound

large amplitude

time →

Louder **volume** → increased **amplitude**.

Adjust the **pitch**:

low-pitched sound

time →

high-pitched sound

time →

Higher **pitch** → increased **frequency**.

Top Tip
We can adjust the amplitude and frequency of all kinds of wave.

Quick Test

1. What sends out signals?
2. What takes in signals?
3. What are the energy changers of a telephone?
4. Give two advantages of a telephone over normal speech.
5. If the amplitude increases, what has changed?
6. If the frequency increases, what has changed?

Optical fibres

Reflection from a plane mirror

When a ray of light strikes a plane mirror it is reflected so that the **angle of incidence** is equal to the **angle of reflection**. The angles are always measured from the normal.

$$\boxed{\text{angle } i = \text{angle } r}$$

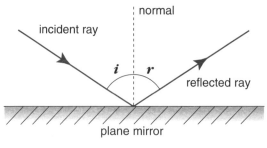

angle i = angle r

Top Tip
Angles are always measured between the ray and the normal. A normal is a line at 90° to the surface.

Total internal reflection

The inner surface of a glass block can sometimes behave like a mirror. This is called **total internal reflection**.

Total internal reflection happens if the ray strikes the inside surface at an angle greater than the **critical angle**.

If the angle of incidence is **small**, the ray is **refracted**.
If the angle is **large**, **total internal reflection takes place**.

Total internal reflection only occurs when light travels from a more optically dense material to a less optically dense material (e.g. glass to air).

Refraction

If angle i is less than the critical angle the ray is refracted and passes out of the glass. A small amount is reflected.

Critical angle

If angle i equals the critical angle c the ray emerges along the edge of the glass block.

Reflection

If angle i is greater than the critical angle the ray is totally internally reflected.

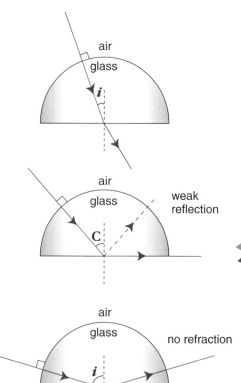

Top Tip
Always put arrowheads on rays. If the ray changes direction, put one arrowhead on the ray before the change and one after.

Optical fibres and their advantages

Optical fibres are very pure, **thin, flexible, glass** rods about 0.1 mm diameter.
They are used to **transmit light**.

Optical fibres have a less dense glass coating around the core.
The light travelling inside the narrow optical fibre always hits the glass boundary at a large angle, **greater than the critical angle**.
The light travels by a series of **total internal reflections**.

An optical fibre can carry a signal from a **transmitter** (T) to a **receiver** (R):

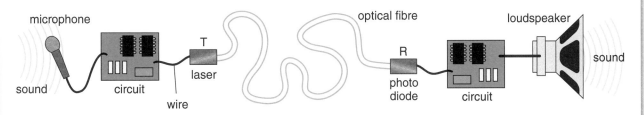

Advantages

- Optical fibres **transmit light at high speed**: 2×10^8 m/s.
- There is **little loss of energy** due to the pure glass.
- Very few repeater units are needed. Copper wires need repeaters every 4 km, optical fibres only every 100 km.
- Signals can be sent over **long distances** optically.
- Optical signals are **free from electrical interference**.
- Optical cables are **much lighter** than electrical copper cables.
- They **cost less**.
- They **carry more signals**.

copper cable

optical fibre

Optical fibres can carry telephone, cable TV, videotext and computer signals into our homes.

Quick Test

1. What is the Law of Reflection?

2. What invention transmits light through an optical fibre?

3. What is the name for light bouncing of the inside wall of the fibre?

4. What is done to stop light escaping from the glass?

5. Give five advantages of an optical fibre over copper cable.

Radio

Radio transmission

All **radio and light waves travel through space at the same speed**:

$$v_{radio} = v_{light} = 300\,000\,000\,\text{m/s} = 3 \times 10^8\,\text{m/s} \qquad v = f\lambda$$

Top Tip
Make sure you understand modulation.

Each radio transmission station broadcasts using its own **wavelength** or **frequency** of radio wave.

The sound is **carried** by the radio wave by **modulation**.

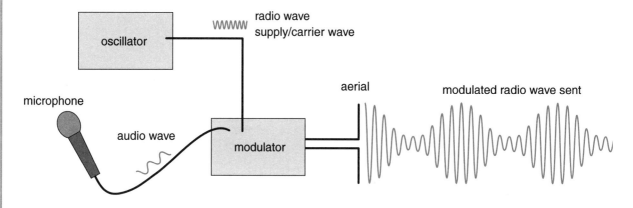

Amplitude Modulation (AM)

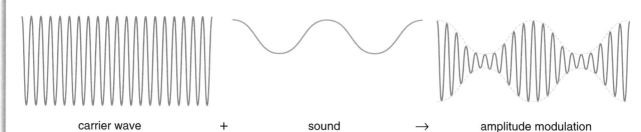

| carrier wave | + | sound | → | amplitude modulation |

Radio reception

The radio

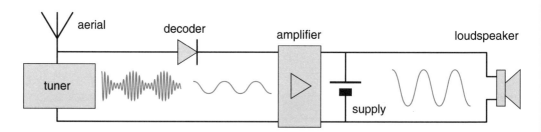

Aerial: **Picks up** many radio waves and changes them to electrical signals.

Tuner: **Selects** one radio frequency.

Decoder: **De-modulates the sound pattern** from the radio wave.

Amplifier: **Increases the amplitude** of the signal.

Supply: **Power** for the amplifier and the bigger signal.

Loudspeaker: **Changes the electrical signals into sound waves**.

Diffraction

Waves can bend (**diffract**) around obstacles. This is seen on the ripple tank:

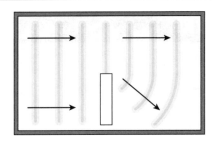 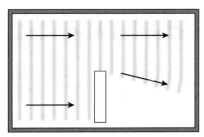

Long waves (LW) diffract more than short waves (SW):

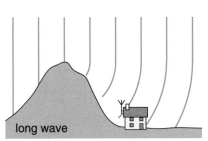 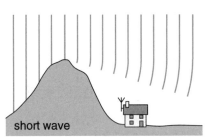

long wave   short wave

Radio waves have a longer wavelength than TV waves.
Radio waves diffract more than TV waves.

The radio spectrum

Top Tip
Diffraction is a property of all waves.

Some radio waves are bounced off the ionosphere

ionosphere

reflection of radio waves

transmitter   receiver

charged particles above the Earth

Radio bands

LW	Long Wave	Good diffraction. Travels far.
MW	Medium Wave	Local and distant broadcasts.
SW	Short Wave	Navigation. Long distance reflection off ionosphere.
VHF	Very High Frequency	Short distance communication. FM sound in stereo.
UHF	Ultra High Frequency	Short distance. Air – air. Air – ground. Colour TV.
SHF	Super High Frequency	Microwave. Satellite communication.

Quick Test

1. At what speed do radio waves travel at?

2. If Radio 1 has a wavelength of 285m, what is its frequency?

3. What name is given to the adding of audio and radio waves?

4. What is the job of the tuned circuit in a radio?

5. Why might a radio be heard when a TV cannot?

Answers 1. 3×10^8 m/s 2. 1.05 MHz 3. Modulation 4. To select one frequency 5. Radio waves diffract more due to longer wavelength.

Television

TV transmission

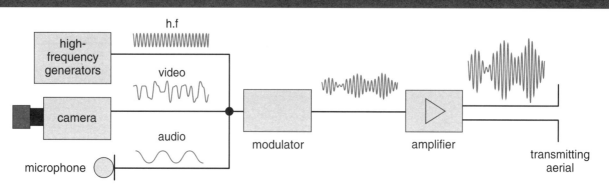

A **TV transmitter** has to send **audio** and **video** signals.

The **modulator** has to **combine the audio and video signal information** onto the high frequency signal.

The **aerial** transmits radio and TV waves in all directions.

TV receiver

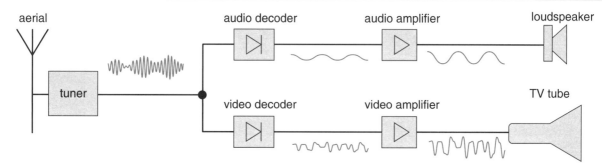

Aerial:	**Picks up** many TV waves, changes them to electrical signals.
Tuner:	**Selects** one frequency.
Video	**Decoder: De-modulates video pattern** from TV wave signal.
Audio	**Decoder: De-modulates audio pattern** from TV wave signal.
Video	**Amplifier: Increases amplitude** of the video signal.
Audio	**Amplifier: Increases amplitude** of the audio signal.
TV Tube:	Changes electrical signal to **light**.
Speaker:	Changes electrical signal to **sound**.

Top Tip
When asked to describe, for example, how a TV works, don't ramble – list the facts.

Vacuum tube

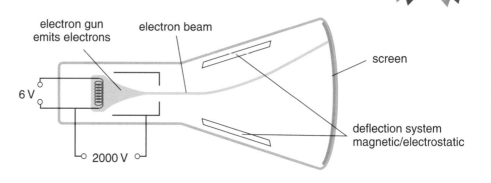

TV images

On Screen
A beam of electrons hits the screen. The **phosphor dots on screen glow and give out light**.

Line build
An image is built up line by line. The **electron beam scans across each line**. 625 line/picture. 25 pictures/s.

Brightness variation
More electrons are fired for a brighter spot, less for dark.

Image retention
A new picture arrives every $\frac{1}{25}$ s.

Our eye can hold an image for $\frac{1}{10}$ s. This is called **persistence of vision**.

Movies
As a new image arrives before the last one goes we see **continuous motion**.

'flyback'

Colour

Top Tip
TV uses colour mixing of light – paint uses different primaries to reflect light – don't get confused.

Colour TV
A colour TV tube has three guns firing electrons at three kinds of spots on the screen. A shadow mask makes sure the beams hit the right spot. The three spots glow **red**, **green** and **blue**.

Colour Mixing
Red, green and blue are called the **primary colours**.

Red light and **blue** light gives **magenta** light.
Blue light and **green** light gives **cyan** light.
Green light and **red** light gives **yellow** light.

Red, **green** and **blue** light gives **white** light.
If all the guns are **off,** the screen is **black**, with no colour.

Magenta, **cyan** and **yellow** are called the **secondary colours**.

Other colours are obtained by varying the intensity of the red, green and blue. To vary the intensity we vary the number of electrons fired.

Quick Test

1. What signals are added to the high frequency TV wave?
2. What part of a TV transmitter combines the waves?
3. What selects the TV station you want to watch?
4. What are the three main parts of a TV tube?
5. Why can we see "movies".
6. What are the three primary colours of light?

Answers 1. Audio and video **2.** Modulator **3.** Tuner **4.** Electron gun, deflection system, screen **5.** Persistence of vision means we can hold images for longer than it takes for the image to be changed **6.** Red, green and blue.

The mobile phone to satellites

Mobile phones and networks

Mobile phones have given us the ability (along with radio & TV) to **communicate without cables**.

The link between the transmitter and receiver is made with **UHF radio waves**.

The mobile networks link into the public networks which use cables, optical fibres and microwaves.

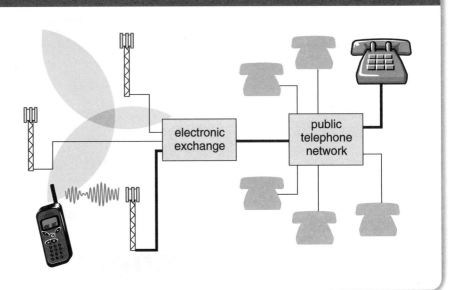

Microwaves

Radio wavebands are becoming crowded.
Above radio frequencies we find **microwaves which have higher frequencies** (in the GHz range).
Microwaves **diffract (bend) so little** though; we say they only **travel in straight lines**.
Repeaters are needed for long distances.

Microwaves are easily **focussed using dish aerials**.

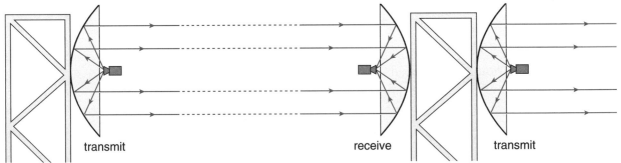

transmit   receive   transmit

The **transmitter** and **receiver** are placed at the **focus point of the dish**.

Larger dishes are used to collect and reflect more rays of energy for a stronger signal.

Microwaves travel at the same speed as light, radio waves and TV signals.

$$v_{micro} = v_{radio} = v_{light} = 300\,000\,000\,\text{m/s} = 3 \times 10^8\,\text{m/s}$$

Microwaves, light, radio and TV are all waves transferring **energy**.

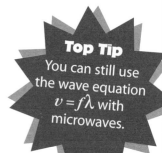

Top Tip
You can still use the wave equation $v = f\lambda$ with microwaves.

Satellites

Three satellites can send a signal around the world.

Microwaves are sent to and from satellites.

The **satellites** act as **repeaters** and **amplifiers**.

The **higher** a satellite is above the earth, the **longer the period of orbit**.

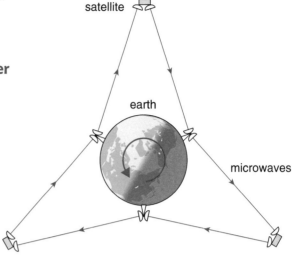

satellite

earth

microwaves

A geo-stationary satellite will be in orbit 36 000 km above the earth.

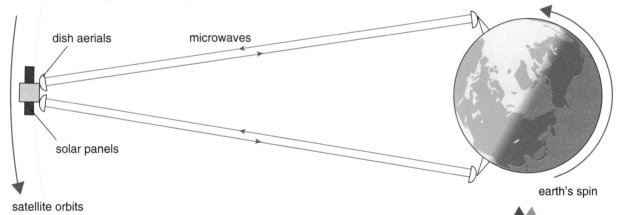

dish aerials

microwaves

solar panels

satellite orbits

earth's spin

A **geo-stationary satellite** appears to be **stationary above the earth** as the earth spins.

The **period of orbit will be 24 hours** – the same period as the earth takes to rotate.

Top Tip
Now is a good time to revise all the different ways to communicate, e.g. by using electricity, light, radio, TV and microwave.

Quick Test

1. Why do mobile phones not need cables?
2. At what point do all the waves from a dish aerial meet?
3. What does a bigger dish do?
4. Why is the period of orbit of a geo-stationary satellite 24 hours?
5. What speed do microwaves travel at?

Answers 1. They use UHF radio waves. 2. Focus point. 3. Catch more rays for a stronger signal. 4. So that it remains above the same point of the earth. 5. 3×10^8 m/s

Test your progress

Use the questions to test your progress.
Check your answers at the back of the book on page 108–109.

1. What do waves carry from place to place?

....

2. Look at the diagram of a wave shown on the right.
What is A? What is B?

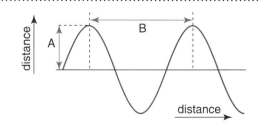

....

....

3. Name three ways in which the direction of a wave can be changed?

....

4. Which reflected ray shows the direction of the
incident ray after striking the mirror,
A, B, C or D?

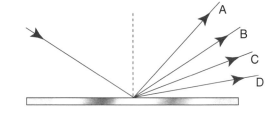

....

....

5. Explain why a flash of lightning is seen before the sound of thunder is heard.

....

6. The diagrams (right) show wave patterns.
Match each of the patterns with the
descriptions A, B, C and D
  a) A low pitched quiet sound
  b) A high pitched loud sound
  c) A low pitched loud sound
  d) A high pitched quiet sound

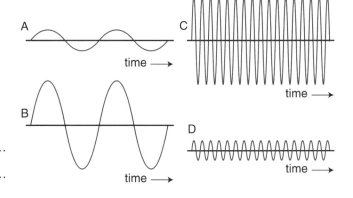

....

....

....

7. What is a device called which sends signals to a satellite?

....

8. What is the speed of electrical signals in telephone wires?

....

9. What type of wave travels between mobile phones?

....

10. What 2 similar devices come after the tuner in a TV?

....

11. What colours are used to produce a TV picture?

....

12. What time does a geo-stationary satellite take to orbit the earth?

....

13. What change would be seen on an oscilloscope when the frequency increases?

....

14. What is an amplifier used for in a radio?

....

15. How does light travel in an optical fibre?

....

16. What part of a radio picks up all the radio signals from the air?

....

17. A wave machine makes 30 waves per minute in a swimming pool. The waves are 3 m long.
A child bobs up and down by 0.6 m on the waves?
a) What is the frequency of the waves?
b) What is the speed of the waves in the pool?
c) What is the amplitude of these waves?

....

18. Why does a curved dish behind a receiver make the signal stronger?

....

19. What is meant by calling a satellite geo-stationary?

....

20. If microwaves have a wavelength of 0.025 m, what is their frequency?

....

21. What is the process called when an audio signal is added to a carrier wave?

....

22. How long will it take for light to travel through 1000 km of optical fibre?

....

23. How is orange produced on a TV screen?

....

24. Which travels faster, light signals in optical fibre or electrical signals in copper wire?

....

25. Why may a house in the hills not receive a local TV signal?

....

At home with energy and power

Electrical energy is one of the most convenient forms of energy. It is easily converted into other forms of energy.

Top Tip
Energy is the key to electricity. The ideas of voltages, power, and circuits are all connected to energy.

Energy changers

Around the home we will find many electrical appliances.

All electrical appliances change electrical energy into other forms of energy.

In the lounge:

TV	electrical → light + sound
Light bulb	electrical → light + heat
Hi-fi	electrical → sound
Heater	electrical → heat + light
Phone	electrical → sound

In the bedroom:

Light bulb	electrical → light + heat
Clock	electrical → kinetic
Radio	electrical → sound
Electric blanket	electrical → heat
Hair dryer	electrical → kinetic + heat

In the bathroom:

Light bulb	electrical → light + heat
Shaver	electrical → kinetic

In the kitchen:

Light bulb	electrical → light + heat
Washing machine	electrical → heat + kinetic
Dryer	electrical → heat + kinetic
Cooker	electrical → heat
Food mixer	electrical → kinetic

In the garden:

Grass cutter	electrical → kinetic
Hedge cutter	electrical → kinetic

DIY and household:

Electric drill	electrical → kinetic
Circular saw	electrical → kinetic
Iron	electrical → heat
Vacuum cleaner	electrical → kinetic

Can you add one of your own to the groups above?

Around the home, the main forms of energy produced are **heat**, **light**, **sound** and **kinetic**.

Power ratings

The power of an appliance is a measure of how quickly energy is supplied.

Power (P) is measured in **Watts (W)**.

Appliances come in a range of powers. For example, an electric drill might be rated from 300 W to 1000 W.

More powerful appliances can do more work or may last longer but cost more to run.

Put each of the electrical appliances in the table into one of the following groups:

- electrical → heat
- electrical → kinetic
- electrical → sound
- electrical → light

Important notice

Appliances which give out heat usually use the most electricity!

Electrical markings

The power rating of most electrical appliances can be found marked on the appliance or a plate attached to it. Remember the unit for power is the Watt (W).

Top Tip

Learn the power ratings of appliances around your home. Understanding the power ratings is vital.

Typical power ratings in Watts:	
Appliance	**Power rating**
Cooker	12000
Washing machine	3000
Tumble dryer	3000
Heater	2500
Kettle	2000
Iron	1900
Vacuum cleaner	1400
Hair dryer	1200
Circular saw	1000
Lawn mower	900
Drill	700
Computer	250
Food mixer	200
Fridge	100
TV	100
Hi-fi	80
Light bulb	60
Radio	20
Clock	10

MODEL AZ1378/29
230V   50Hz ~
22W   C€

Other data that will be found marked are: Voltage e.g. 230 V and frequency e.g. 50 Hz. More about these later.

Quick Test

1. What are the main energy transformations in household appliances?

2. Name three high power appliances.

3. What will a low power appliance not be giving out?

4. Where would you find the power of an appliance?

5. What would be the power of a table lamp?

Answers 1. electrical → heat, electrical → kinetic, electrical → sound, electrical → light **2.** cooker, heater, iron **3.** heat **4.** Marked on it or a plate attached to it **5.** 60W – light bulb

At home with flexes, fuses and plugs

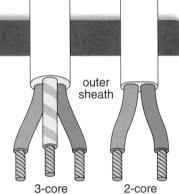

Flexes

The **current** from the supply to an appliance depends on the **power** rating of the appliance. This current travels in a flex (flexible cord) which will have 2 or 3 cores.

Appliances with a high power rating require a thick flex to carry the large currents. If the flex is too thin for the current, the wire can overheat, creating a fire hazard.

Appliance maximum power (W)	Flex rating (A)	Flex area (mm²)
690	3.0	0.5
1380	6.0	0.75
2300	10.0	1.00
3100	13.5	1.25
3680	16.0	1.50

outer sheath

3-core   2-core

Fuses

Fuses are fitted to plugs to protect flexes and appliances. When a fuse blows (melts), no current can exist in the flex.

In domestic situations, there are two ratings for fuses, depending on the power of the appliance.
- The plugs of low power appliances (<700W), which draw low currents, are fitted with a **3A fuse**.
- The plugs of high power appliances (>700W), which draw larger currents, are fitted with a **13A fuse**.

If the fuse used has too high a rating, it will not protect the flex when too large a current occurs. The flex will overheat and this would be a fire hazard.

Fuses are given a **rating** which indicates the **maximum current** that can pass through it without it melting.

The most common fuses in the UK have ratings of **1A**, **3A**, **5A**, and **13A**.

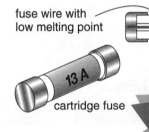

fuse wire with low melting point

13 A

cartridge fuse

Top Tip
If you fit a plug at home, check it has the correct fuse – it could save your life!

The three-pin plug

The voltage of the electricity from cells and batteries is quite low e.g. 9V, 12V. The voltage from the mains is about 230V, which can kill a person. **It can be dangerous** if not used safely.

Most appliances are therefore connected to the mains using **insulated plugs**. It is very important that the wires are connected to the **correct pins**.

Looking at an open plug like the one shown here, the **BR**own wire goes to the **B**ottom **R**ight and the **BL**ue wire goes to the **B**ottom **L**eft. The green and yellow wire goes to the pin at the top.

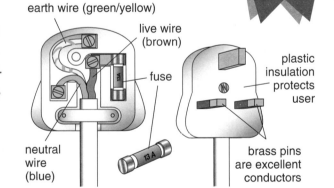

earth wire (green/yellow)

live wire (brown)

fuse

plastic insulation protects user

neutral wire (blue)

brass pins are excellent conductors

Remember, the human body is a conductor of electricity and moisture increases its ability to conduct. Never remove a plug with wet hands.

Earthing and safety

The three pin plug usually has three wires connected to it.
- The **live** wire carries the electrical energy to the appliance.
- The **neutral** wire completes the circuit for the current.
- The **earth** wire is a safety device.

How does the earth wire protect us?
- The earth wire is connected to the metal casing of an appliance.
- A live wire becomes loose and touches the casing.
- A momentary large surge of current is drawn from the live and runs to earth.
- The large surge in current immediately **blows the fuse** in the live wire.
- The supply is cut off – there is no current to run through us.

fuse blows

loose live wire

What happens if the live and neutral are reversed in the plug?
- The internal switch and fuse are no longer onto the live.
- The switch and fuse are now onto the neutral.
- If the appliance's switch is off or its fuse blows – the circuit is still broken, but ...
- The appliance is **off** but is still **live**.
- Touch the appliance and current will run through you to earth!

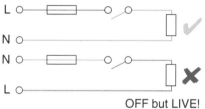

OFF but LIVE!

Remember
- The plug fuse protects the flex from overheating with too much current.
- The earth wire, in a fault, causes a large surge in current, which blows the fuse.
- The switch must be connected to the live wire, so if the switch is off, the appliance is not live.
- The fuse must be connected to the live wire, so if the fuse blows, the appliance is not live.

Danger
- Keep electrical appliances away from water.
- Always use the correct fuse.
- Never use wrong, frayed or badly connected flexes.
- Never overload multiway adaptors.

Double insulation

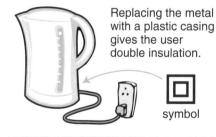

Replacing the metal with a plastic casing gives the user double insulation.

symbol

Appliances such as electric kettles and drills now have plastic casings. These casings cannot become live. These appliances are said to be double insulated. These appliances carry the double insulation symbol. Double insulated appliances do not require an earth wire.

Top Tip
Try to find appliances with the double insulation symbol and 2-core flex at home.

Quick Test

1. What happens to a flex if it is too thin?

2. What size of fuse would you fit to a **a)** 500 W and **b)** 2000 W appliance?

3. What is the colour and position of the wires in a 3 pin plug?

4. State three causes of electrical accidents

5. Draw the double insulation symbol.

AC/DC

There are various types of electricity: static, d.c. and a.c.

Static electricity

Did you know that lightning is caused by static electricity?

Two insulators rubbing can become electrically charged by the transfer of electrons (negative charges) between them.

The object that **gains electrons** is said to be **negatively** charged. The object that **loses electrons** is said to be **positively** charged.

Like charges **repel**. **Unlike** charges **attract**.

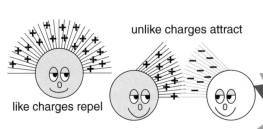

uncharged plastic rod

rod gains electrons and becomes negatively charged

uncharged cloth

cloth loses electrons and becomes positively charged

unlike charges attract

like charges repel

Top Tip
The C.R.O. measures voltage on the y-axis and time on the x-axis. Learn how to change and read these scales for your practical assessment.

Battery or mains?

We can study these supplies using a Cathode Ray Oscilloscope.

The horizontal line shows the voltage is steady and current is steady and in **one direction**. A battery supplies d.c. (**direct current**).

voltage
0
time →

The wave shows the voltage varies (0 to peak). The current **varies** in **size** and **direction**. The mains supplies a.c. (**alternating current**).

voltage
0
time →

The frequency of the mains supply is **50 Hz**. This means there are 50 complete waves per second.

The declared value of the mains is **230 V**. This is less than the peak value. (~ 70% of peak in fact).

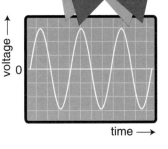

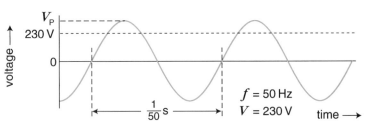

voltage
V_P
230 V
0
$\frac{1}{50}$ s
$f = 50$ Hz
$V = 230$ V
time →

Circuits

Charges flow in a complete circuit. The components in circuits are drawn using symbols:

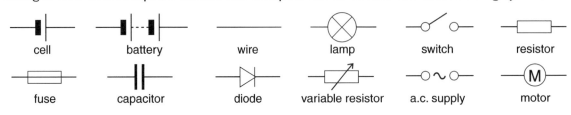

| cell | battery | wire | lamp | switch | resistor |

| fuse | capacitor | diode | variable resistor | a.c. supply | motor |

Current

Electrons are **free to move in a conductor** e.g. a metal.
Most non-metals are insulators as the electrons are not free to move.
When **charges flow** we have created an **electric current**.

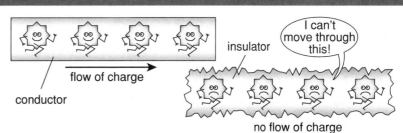

flow of charge

conductor

insulator

I can't move through this!

no flow of charge

We **measure current** with an **ammeter**.
The size of a current is the **rate at which charge is flowing**.
Charge (Q) is measured in **Coulombs (C)**.

Current (I) is measured in **amperes (A)**.

If a current of 3A passes through the bulb, 3C of charge flows through it each second.

$$I = \frac{Q}{t} \qquad Q = It$$

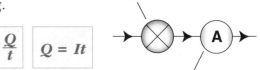

bulb

ammeter measuring the current through the bulb

Voltage

- As charges flow around a circuit they **give away the energy** gained from the cell or battery.
- This **energy is transferred into other forms** by the components in the circuit.
- A bulb **transfers** electrical energy into heat and light energy.
- A resistor **transfers** electrical energy into heat energy.
- A buzzer **transfers** electrical energy into sound energy.

How much energy is transferred?

- The **voltage** or **potential difference (pd)** tells us **how much energy is transferred** by a component in a circuit.
- A **voltage of 1 V** means **1 J of electrical energy** is being transferred into other forms **every time 1 C** of charge passes through a component.
- We measure the voltage across a component using a **voltmeter**.
- This voltmeter (right) is measuring a voltage of 4 V across the bulb.
- The bulb is changing **4 J of electrical energy** into **4 J of heat and light energy** every time **1 C of charge passes through the bulb**.

Top Tip
You will need to learn how to correctly place an ammeter and voltmeter for your practical assessment.

This bulb is changing 4 J of electrical energy into 4 J of heat and light energy every time 1 C of charge passes through it.

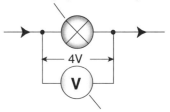

4V

Voltmeter, measuring the voltage across the bulb.

Quick Test

1. What is an electric current?

2. In what units do we measure:
   a) charge?   b) current?
   c) energy?   d) voltage?

3. What instrument do we use to measure
   a) current?   b) voltage?

4. What is a battery?

5. Explain why metals are good conductors of electricity and non-metals are mainly poor conductors.

Electrical resistance and power

Resistance

Using resistors

We can use resistors to control the size of the current in a circuit. If a variable resistor is included in a circuit its value can be altered so that the current around the circuit can easily be changed.

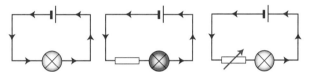

With no resistors in the circuit, there will be a large current.

With a resistor in the circuit, there will be a smaller current.

Altering the value of this variable resistor changes the brightness of the bulb.

The greater the resistance, the smaller the current that exists.

Resistance of a piece of wire

The resistance of a piece of wire depends upon:
- length – the **longer the wire**, the **greater the resistance**;
- thickness – the **larger the cross-sectional area** of a wire, the **smaller its resistance**;

large R small R

- the **material** from which the wire is made, e.g. copper wires have a low resistance and so are often used as connecting wires.

Electrical components

Electrical components in a circuit also resist current. They have resistance.

Resistance (R) is measured in **ohms (Ω)**.

If, when a **voltage** of **1V** is applied across a component, there is a **current of 1A**; **the component has a resistance of 1Ω**.

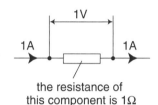

the resistance of this component is 1Ω

The relationship between **resistance**, **current** and **voltage** is: $R = \dfrac{V}{I}$

Example: A current of 3A is created in a circuit when a voltage of 12V is applied across a wire. What is the resistance of this wire?

$$R = \frac{V}{I} = \frac{12}{3} = 4\,\Omega.$$

Resistance can be measured using an **ohmmeter** or using $R = \dfrac{V}{I}$

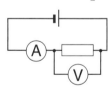

Ohm's law

A range of pds is applied across a conductor and the corresponding currents measured.

Assuming the temperature of the conductor does not change, the graph shows **a straight line passing through the origin**. The shape of the graph shows **the current is directly proportional to the voltage applied.**

This conclusion is known as **Ohm's law**.

$\dfrac{V}{I}$ remains constant for different currents. $R = \dfrac{V}{I}$ $V = IR$ Ohm's Law

current / voltage

Electrical power

Power is a measure of **energy** changed in **unit time**. Power (**P**) is measured in **Watts (W)**.

If a light bulb has a **power rating** of 40 W, it changes 40 J of electrical energy into heat and light energy every second.

If an electrical fire has a **power rating** of 2 kW (2000 W), it changes 2000 J of **electrical energy** into 2000 J of heat and light energy **every second**.

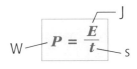

$$P = \frac{E}{t}$$

How many joules of energy have been converted?

To calculate the total amount of energy an appliance has converted we use the equation:

$$\text{Energy} = \text{Power} \times \text{time}$$
$$\text{or} \quad E = P \times t$$

Example: How much electrical energy is converted into heat and light energy when a 60 W bulb is turned on for 5 minutes?

$$E = P \times t = 60 \times 300 = 18\,000 \text{ J or 18k J.}$$

More power equations

A 12 V, 24 W lamp draws a current of 2 A.
A 12 V, 36 W lamp draws a current of 3 A.

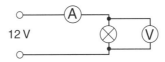

12 V

The product VI calculates the energy transformed each second – the power $\boxed{P = VI}$

Now combine this with Ohm's law: (We will now know four equations for power).

$$P = IV = I(IR) \text{ so: } \boxed{P = I^2 R} \quad \text{and } P = VI = V\frac{V}{R} \quad \text{so} \boxed{P = \frac{V^2}{R}}$$

Top Tip
Power equations:
$$I = \frac{P}{V}$$
can be used to select a fuse value.

Lamps and heaters

In a lamp, electrical energy is transformed to **light** and **heat**.
In a **filament lamp** this takes place in a **resistance wire**.
In a **discharge lamp**, this takes place in the **gas**.
The discharge lamp is **more efficient** than the filament lamp as **more energy is transformed into light**.

An **electric heater** also has a resistance wire known as the **element**. More energy goes to heat than light.

Quick Test

1. Give three practical uses of variable resistors.

2. A circuit resistance of 24 Ω is changed to 12 Ω. What will happen to the current?

3. A 6 Ω resistor is attached to a 24 V supply. What current is drawn?

4. A 2300 W heater is attached to the mains. What current exists?

5. A Joule meter records 2400 J of energy being supplied to a low voltage heater in 1 minute. What is the power of this heater?

6. Where does the energy transformation take place in a heater?

Voltage and current in circuits

The two main ways of connecting components in a circuit are in **series** or in **parallel**.

Series circuits

Series circuits have all the components in a row or loop. There are no branches.
There is only one path for the flow of charge.
A series circuit is turned on or off by a single switch or break anywhere in the circuit.

The current is the same at all points of a series circuit.

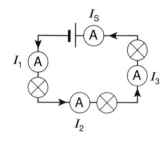

$$I_S = I_1 = I_2 = I_3$$

The sum of the voltages across all the components in series is equal to the supply voltage.

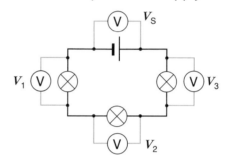

$$V_S = V_1 + V_2 + V_3$$

Top Tip
Learn how voltage and current behave in series and parallel circuits.

Parallel circuits

Parallel circuits have branches and junctions.
There is more than one path for the charges to follow.
A break in one branch has no effect on the other branches.
Switches can be put in to turn off all or part of the circuit.

The sum of the currents in parallel branches is equal to the current drawn from the supply.

circuit has branches and more than one path to follow

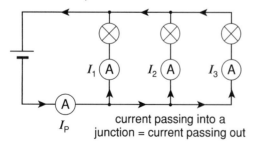

current passing into a
junction = current passing out

$$I_P = I_1 + I_2 + I_3$$

The voltages across components in parallel are the same and equal to their supply.

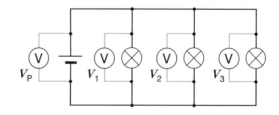

$$V_P = V_1 = V_2 = V_3$$

Take care with parallel circuits. Adding too many appliances to one socket is dangerous as too large a current could be drawn from the supply. Overheating may occur.

2-way circuits

Stair lighting uses two or more 2-way switches in series.

Having gone upstairs, you do not need to come back down to switch the lights out!

This circuit is in the **off** position.

Either switch will turn the light **on**.

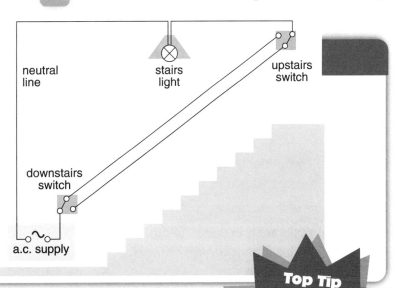

Top Tip
Try to understand why switches are sometimes connected in series and sometimes in parallel.

Car lighting circuits

A car has special requirements:

- If a bulb blows, the rest of that circuit must stay on.
- The sidelights should come on first and together.
- The headlights come on together after the sidelights are on.
- Also note, a car often uses the car chassis as a 'return' instead of wire.
- The positive is the 'live' and the negative is attached to the chassis for the 'return'.

Can you see how this circuit would work?

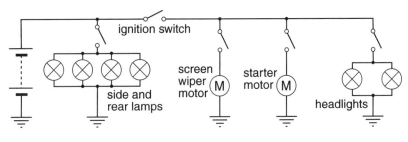

All switches must be closed when switches are in series.

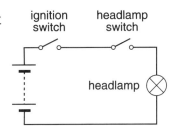

Either switch can be closed (opening door) to operate this light. The switches are in parallel.

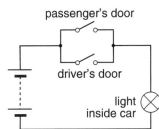

Quick Test

1. When components are in a row, this is a .... circuit

2. In which type of circuit:
   a) is the current the same at all points?
   b) can just part of the circuit be turned off?
   c) are there branches or junctions?

3. If a 6 V battery is placed across three identical bulbs in series, what is the middle bulb's voltage?

4. If the voltage across a bulb is 3 V when connected in parallel with another identical bulb, what is the supply voltage to these bulbs?

5. In a car, the headlamp switch cannot be closed before the .... switch.

Resistance in circuits

Resistors in series and parallel

Series

If we join components **in series** we **increase the resistance** of the circuit.

The current will **decrease**.

The **total resistance in series is equal to the sum of the individual resistances**.

$$R_S = R_1 + R_2 + R_3$$

Parallel

If we join components **in parallel** we **decrease the resistance** of the circuit.

The current will **increase**.

The combined resistance in parallel is calculated using a less than straightforward formula!

$$\frac{1}{R_P} = \frac{1}{R_1} + \frac{1}{R_2} + \frac{1}{R_3}$$

For two resistors in parallel this becomes: $R_P = \dfrac{\text{product}}{\text{sum}}$

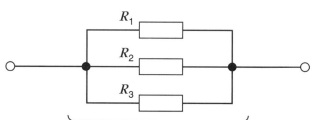

Top Tip
The combined resistance of parallel branches is always smaller than the smallest resistor!

Resistor combinations

Calculate a combination circuit by combining the parallel resistances first, then add the series resistances to find the total.

$$\frac{1}{R_P} = \frac{1}{R_1} + \frac{1}{R_2} = \frac{1}{4} + \frac{1}{4}$$
$$\Rightarrow R_P = 2\,\Omega$$

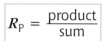

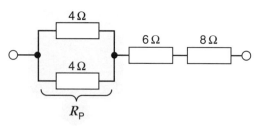

so $R_S = R_P + R_1 + R_2$
$= 2\,\Omega + 6\,\Omega + 8\,\Omega$
$= 16\,\Omega$

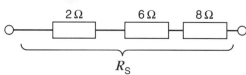

Top Tip
Two identical resistors in parallel have half the resistance of one!

Overloading

Adding more and more branches to a parallel circuit will draw more and more current from the supply as the resistance decreases. This is dangerous as it could lead to overheating of the socket.

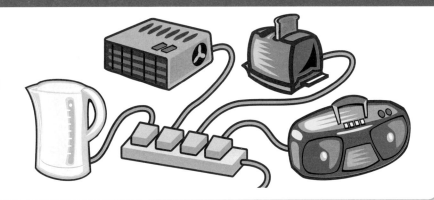

Fault finding

Two very common electrical faults:

open circuit

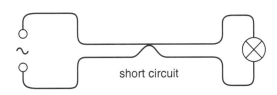

short circuit

Open circuit – A loose or broken wire. The open circuit resistance on an ohmmeter shows $\infty\,\Omega$.

Short circuit – Two bare wires touch or are crossed by loose wire. Short circuit resistance = $0\,\Omega$.

A simple continuity tester

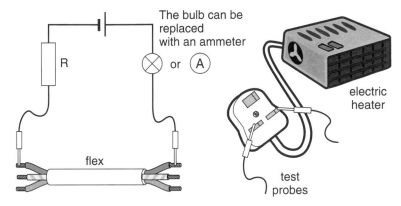

The bulb can be replaced with an ammeter

R

$\otimes$ or A

flex

electric heater

test probes

R is a limiting resistor for current.

When the test probes are joined R protects the bulb from too much current. The bulb should light!

The circuit can check the wires in the cable separately. If any are broken the bulb will not light.

Quick Test

1. If resistance increases, what happens to the current?

2. If 2, 5 and 7 ohm resistors are joined in series, what is the total effect?

3. A $4\,\Omega$ and a $6\,\Omega$ resistor are joined in parallel. What is the combined resistance?

4. You have four $10\,\Omega$ resistors. How do you obtain a total resistance of $25\,\Omega$?

5. A ohmmeter reading shows $0\,\Omega$ when testing an empty lamp-holder. Is anything wrong?

Answers 1. Decreases **2.** $14\,\Omega$ **3.** $2.4\,\Omega$ **4.** Two resistors in parallel = $5\,\Omega$, then add two resistors in series. **5.** Yes. The reading should show 'infinity'. The lamp-holder has been short circuited

House wiring

Circuits

A few items such as an electric cooker, shower or immersion heater are high power and have an individual cable containing live (brown), neutral (blue) and earth (green/yellow) wires.

Household wiring connects appliances in parallel.
In this way we can switch lights and appliances on and off without affecting other lights or appliances.
The lighting and power circuits in a house are types of parallel circuit.

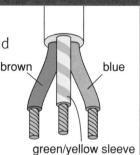

brown   blue

green/yellow sleeve

Lighting circuit

House lights are connected in parallel. There may be an upstairs circuit and a downstairs circuit.

The cable carries live, neutral and earth but is thinner and cheaper than the power circuit.

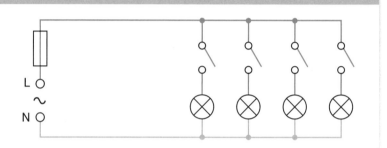

Power circuit

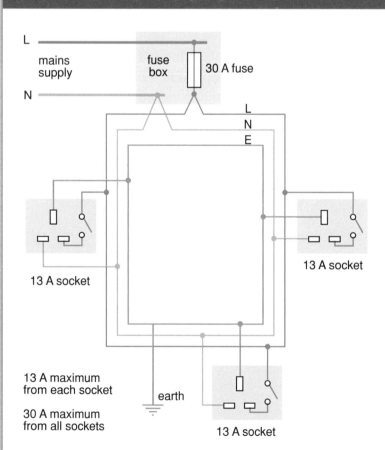

L

mains supply

N

fuse box

30 A fuse

L
N
E

13 A socket

13 A socket

13 A maximum from each socket

30 A maximum from all sockets

earth

13 A socket

The power circuit is a special type of parallel circuit. It is called a **ring** circuit.

Live, neutral and earth wires are each in a continuous loop connected to the **consumer unit supply**.

There are two paths for current to and from each socket from the supply. Half the current flows each way. This allows thinner cable to be used, as the circuit can take twice the current that one path cable would allow.

The cable is still thicker than the lighting circuit cable, carries more current and handles more power.

Top Tip
The ring circuit is a parallel circuit. The branches are between the live and neutral wires.

Fuses and circuit breakers

The **consumer unit** contains **mains fuses** or **miniature circuit breakers**, which protect the mains wiring.

The consumer unit is often found near the house entrance beside the meter:

Consumer Unit Fuses:

Lighting circuit	5 A
Ring circuit	30 A
Immersion heater	20 A
Cooker	45 A

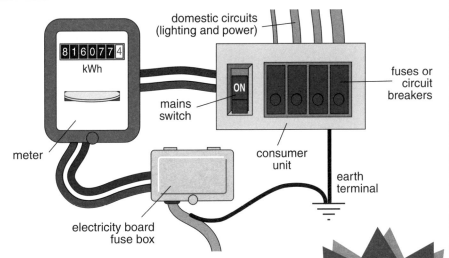

domestic circuits (lighting and power)

fuses or circuit breakers

mains switch

consumer unit

earth terminal

meter

electricity board fuse box

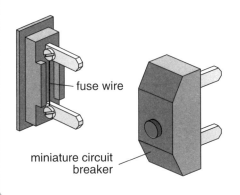

fuse wire

miniature circuit breaker

A mains fuse will 'blow' or circuit breaker 'trip' if the circuit is overloaded.

A **miniature circuit breaker** (**MCB**) is an **automatic switch**, which can be used instead of a fuse.

After a fault has been corrected, a MCB can be **reset** instead of being replaced like a burnt out fuse.

Top Tip
Ask to be shown the consumer unit in your home. Check the type of circuits and MCB values if visible or listed.

Watt's the cost

The electricity meter records the number of 'units' of electrical energy used.

This unit of energy is the **kilowatt-hour**.

$$E = P \times t$$

but power must be in kilowatts and time in hours to find the number of 'units'.

Example: Calculate the energy used when a 3kW fire is turned on for two hours. The cost of each unit is 15p.

$E = P \times t = 3 \times 2 = 6$ kWh or 6 units.

Cost: $6 \times 15 = 90p = £0.90$

Quick Test

1. Why are household appliances connected in parallel?

2. What are the advantages of using the ring circuit as the preferred method of wiring in parallel?

3. What is the modern replacement for the fuse?

4. How many joules of energy are there in a kilowatt-hour?

5. At a cost of 15p / unit, what is the cost of
   a) a 60W bulb on for 10 hours?
   b) a 3000W fire turned on for 3 hours?
   c) a 2000W tumble dryer used for 1h 30mins.?

Answers 1. They can be switched individually **2.** Two paths for current, thinner cable is less costly **3.** MCB **4.** $E = P \times t = 1000 \times (60 \times 60) = 3\,600\,000$ J **5. a)** 0.6 units costs 9p **b)** 9 units costs 135p = £1.35 **c)** 3 units costs 45p

The motor effect

Magnets and electromagnets

All magnets have two poles: North and South.

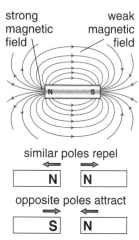

similar poles repel

| N | | N |

opposite poles attract

| S | | N |

Oersted, 1820, discovered that a magnetic field exists around a current-carrying wire.

To **make the magnetic field stronger** we can **increase the current** or make the wire into a **coil** (**solenoid**).

If a solenoid is wrapped onto an **iron core** the **magnetic field strength increases**. This is an **electromagnet**.

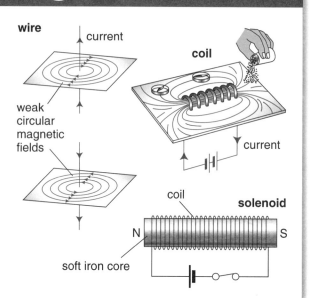

Uses of electromagnets

Electric bell

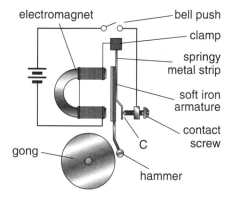

When the bell push is pressed the circuit is complete and the electromagnet is turned on.

The **soft iron armature** is pulled towards the electromagnet and the **hammer** hits the **gong**.

At the same time a gap is created at C and the electromagnet is turned off.

The armature now springs back to its original position and the whole process starts again.

As long as the bell push is pressed, the armature will vibrate back and forth striking the gong.

Relay switch

This is a safety device. It is often used to **turn on a circuit** through which a **large (potentially dangerous) current** passes using a circuit through which a **small current passes**.

When the switch S is closed, a small current flows, turning the electromagnet on. The **rocker** is pulled down and at the same time **the contacts at C** are pushed together. A large current now passes through the second circuit.

When S is opened the **electromagnet is turned off**. The **rocker is released** and returns to its original position. The **contacts at C open** and **current ceases to pass through** the second circuit.

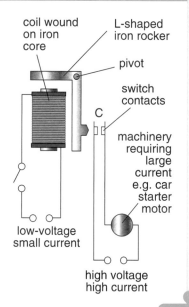

Top Tip
Don't try to remember the circuits for the electric bell, the car breaker or the relay switch. Just remember these as practical examples which make use of the magnetic effect of a current.

Scrap-yard electromagnet

When current passes through the coil, a very strong electromagnet is created which is able to pick up cars.

When the magnet is **turned off** the **magnetic field collapses** and the car is **released**.

The motor effect

We can create movement from electricity.

A **current-carrying wire** experiences a **force** when the **wire is in a magnetic field**. The **direction of the force** on a current-carrying wire depends upon the **direction of the current and of the field**.

force pushes wire downwards

switch

The d.c. motor

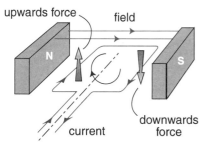

upwards force
field
N
S
current
downwards force

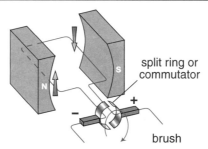

S
split ring or commutator
N
− +
brush

The force on one side will push up. The force on the other side will push down.

The force changes every half turn. The commutator changes the direction of the current in the coil every half turn. The coil will keep rotating in the same direction.

Top Tip
Make sure you know how to change: a) the direction of the force on a wire and b) the speed of rotation of a motor.

To make a motor turn **quicker**, increase:
1. the **size of current**   2. the **number of turns** of coil   3. the **magnetic field strength**

Commercial motors

Carbon brushes are cheap, good conductors. They wear to the commutator's shape, without wearing the commutator. They are easily replaced if they wear away completely.

Multi-section commutators are used with multiple coils. This makes the motor smoother and more powerful. One of the coils will be near right angles to the magnetic field for the greatest turning force. These are the rotating coils.

Stationary field coils are used instead of permanent magnets. This makes the motor lighter and cheaper.

Quick Test

1. State three practical uses of the magnetic effect of current.
2. What is the energy change in a motor?
3. How can you reverse a motor?
4. How can you speed up a motor?
5. What are the field coils?

Answers 1. bell, relay, motor **2.** electrical to kinetic **3.** reverse the current or the field **4.** increase the current, the number of turns of coil, the magnetic field strength **5.** electromagnets

Test your progress

Use the questions to test your progress.
Check your answers at the back of the book on pages 108–109.

1. What happens if two similarly charged objects are placed next to each other?

   ....

2. What happens if the North pole of one magnet is placed next to the South pole of another magnet?

   ....

3. Name the three particles contained in an atom. Which of these particles moves when current is passed through a wire?

   ....

4. Give one advantage that electromagnets have over permanent magnets.

   ....

5. a) Why will charge flow in circuit A but not in circuit B?
   b) Explain how circuit B could be used to test materials to see if they are conductors or insulators.

   circuit A   circuit B

   ....

   ....

6. To which part of an electrical appliance should the earth wire be attached?

   ....

7. What kind of electrical circuit contains different paths for currents to follow?

   ....

8. State three ways in which the speed of rotation of an electric motor could be increased. What is the purpose of the split ring in an electric motor?

   ....

   ....

9. What is the voltage across bulb A in the circuit?

   ....   9 V   12 V

   ....   A

10. Calculate the electrical energy in units when a 2 kW fire is turned on for three hours. Calculate the cost of this energy if the cost of one unit is 11 pence.

    ....

11. Explain the difference between a direct current (d.c.) and an alternating current (a.c.).

    ....

12. Calculate the power of a hair drier which, when connected to a 230 V supply, draws a current of 5 A.

    ....

13. Calculate the resistance of a piece of wire which, when a voltage of 6 V is applied across its ends, draws a current of 0.25 A.

....

14. When a 3 kW fire is connected to an a.c. supply there is a current of 13 A. Calculate the voltage of the a.c. supply.

....

15. Calculate the correct fuse that should be included in a three-pin plug for a 1800 W 230 V hair drier.

....

16. An electric saw shows the double insulation symbol. How many wires are there in the connecting flex?

....

17. A filament lamp and a fluorescent tube are both rated at 40 W. What difference would there be in the amount of light produced and why?

....

18. Give a use for a variable resistor at home.

....

19. A 24 W lamp is switched on for 5 minutes. How much energy was emitted?

....

20. Bulbs for a car are rated at 12 V, 5 W for the sidelights and 12 V, 21 W for the headlights.
   a) What current will be drawn from the battery when the sidelights and headlights are on?
   b) Which bulb has the lower resistance and why?

....

....

21. A current-carrying wire is in a magnetic field. It is experiencing an upwards force.
   a) What happens if the direction of the magnetic field is reversed?
   b) What happens if the direction of current is also reversed?

....

....

22. In commercial motors:
   a) Why are several rotating coils used?
   b) Why are field coils used?

....

....

23. A 60 W lamp is connected to a 230 V supply.
   a) Calculate the resistance of this lamp.
   b) Two such lamps are connected in parallel. What is their combined resistance?

....

....

The use of thermometers

Temperature and thermometers

Temperature tells us how hot a body is and is measured in degrees Celsius (°C).

Some typical temperatures:

North Pole	~	−60°C
Melting ice	=	0°C
Room temperature	~	20°C
Core body temperature	=	37°C
Hot tap water	~	50°C
Boiling water	=	100°C
Sun	~	10^6°C

Anders Celsius, 1742, a Swedish physicist and astronomer, used the two fixed points of melting ice and boiling water to make a temperature scale with 100 divisions.

All thermometers have some property which changes with temperature:

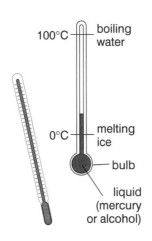

warm   cold

core   core

shell

37°C
35°C
33°C
31°C
29°C

Liquid-in-glass thermometers

The liquid **expands** (more than the glass it is in) when heated.

The liquid **contracts** when cooled.

100°C — boiling water

0°C — melting ice

— bulb

— liquid (mercury or alcohol)

Liquid crystal strip thermometer

Different crystals melt at different temperatures.

Colour change shows the temperatures.

35 36 **37** 38 39 40 41 42

Rotary thermometer

Based on the **bimetallic** strip. Outside metal expands more when heated, causing the strip to bend and a pointer to move.

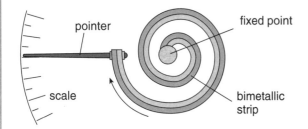

pointer   fixed point

scale   bimetallic strip

Digital thermometers

An electrical property, such as resistance, changes with temperature. The display is digital.

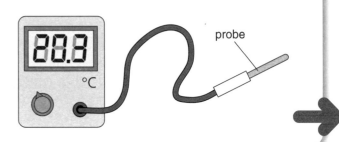

probe

°C

Clinical thermometers

Clinical thermometers need to be able to **hold** the highest reading taken after the thermometer has been removed from the patient. A liquid-in-glass thermometer will have a 'kink' to prevent the liquid falling.

A digital thermometer will hold this reading electronically.

The **range** of the clinical thermometer will be limited and is normally no more than ±5°C around normal body temperature of 37°C, e.g. from 32°C to 42°C.

The clinical thermometer will have **small divisions** of 0.1°C making it sensitive to small changes.

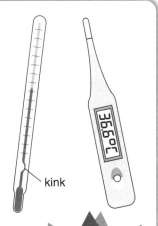

kink

Measuring body temperature

- Shake or reset the thermometer.
- Place under the tongue to measure core temperature.
- Leave for a few minutes for the temperature to rise.
- Remove and read the scale. Note how the reading is held.

Diagnosis of Illness

Our inner **core body temperature** does not normally change from **37°C**.
The core body temperature is for the organs such as the heart, lungs and brain.
The surface of our skin may be cooler, e.g. 33°C.

If we are **too hot** we suffer from **hyperthermia** and are ill!
If we are **too cold** we suffer from **hypothermia** and are ill!

Shivering helps you heat up!

A high temperature helps fight infection!

Temperature (°C)	Condition
>43	Death
41	Convulsions
39	Dilation of blood vessels, heart rate up
38	Fever
37	**Normal**
35	Shivering
34	Constriction of blood vessels, heart rate down
33	Amnesia
28	Loss of consciousness
26–28	Death

Quick Test

1. What is our normal body temperature?
2. How would a nurse recognise hypothermia?
3. Name 3 properties of thermometers which change with temperature.
4. How do you take a patient's temperature?
5. Why is the range of a clinical thermometer limited?

Answers 1. 37°C **2.** temperature drops < 37°C **3.** expansion, crystals melt, resistance **4.** see steps above **5.** to allow fine divisions & too ill beyond range!

Using sound

Vibration of particles

Sounds are produced by **vibrations**.

Sounds need particles to travel through.

Particles vibrate as the energy passes through them.

Sounds can travel **through solids**, **liquids** or **gas** but **not a vacuum**.

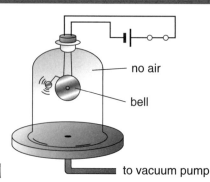

no air

bell

to vacuum pump

no air – no sound

The speed of sound

Sound travels **fastest** in **solids** and **slowest** in **gases**.

Some useful values:

Remember: $v = \frac{d}{t}$ and $v = f\lambda$

Material	Steel	Bone	Blood	Water	Air
Speed of sound (m/s)	6000	4000	1570	1500	340

The stethoscope

Sounds are made in the body at different frequencies.

Lungs – high frequency sound – closed bell with diaphragm.

Heart – low frequency sounds – open bell.

Sounds picked up at the chestpiece are transmitted up through the air in the tubing to the earpieces.

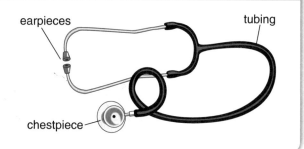

earpieces

tubing

chestpiece

Range of hearing

The range of hearing for a young person is 20Hz–20000Hz (20kHz), decreasing with age.

signal generator

400

loudspeaker

Ultrasound

High frequency vibrations, greater than we can hear, are called **ultrasonic vibrations** or **ultrasounds**.

Ultrasounds range from 20kHz up to MHz. (Dogs, bats and dolphins can hear some of this range.)

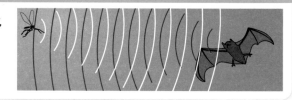

Ultrasonic scanning

Very high frequencies of several megahertz (MHz) are used in medicine to produce images of the inside of the body and in particular of foetuses.

The speed of ultrasound in soft tissue, $v_{tissue} = \sim 1500$ m/s.

The waves are **emitted by a probe** and **reflected by tissues and bones** of the foetus from different depths, so a 3-D picture can be built up.

Jelly or oil is placed between the probe and the skin. This excludes the air to prevent the ultrasound being reflected at the change from air to tissue. The waves would then not enter the mother's body.

The **reflected** waves are detected and the **patterns of reflection times** are used to build up an image on a computer monitor.

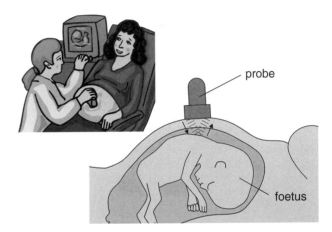

Ultrasounds are far better than X-rays as they do not expose the foetus to harmful radiation.

Ultrasounds are also used in medicine for:

- echograms of the brain
- detecting cavities in teeth
- breaking up kidney stones

as well as other uses outside of medicine.

probe

foetus

Sound levels

Top Tip
Learn some examples of sound levels.

Excessive noise can damage hearing. We measure **loudness of sound** on the **decibel (dB)** scale using a sound level meter.

Sound above 90dB is considered **dangerous**. Ear muffs should be worn to absorb the energy of loud sounds and protect the ears. Loud energy peaks can be more dangerous e.g. lorry, drill or factory sounds. Sound insulation can also absorb sound coming into our homes. Normal conversation is about 60dB.

| 20 | 30 | 40 | 50 | 60 | 70 | 80 | 90 | 100 | 120 | 130 | 140 | dB |

Quick Test

1. What are the three main parts of a stethoscope?

2. What is meant by ultrasound?

3. Why is ultrasound safer than X-rays?

4. At what speed would ultrasound travel in soft tissue?

5. What is meant by noise pollution?

6. What might you expect the noise level at a disco to be?

Light and sight

Refraction of light

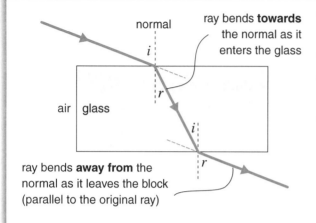

ray bends **towards** the normal as it enters the glass

ray bends **away from** the normal as it leaves the block (parallel to the original ray)

Angles are always measured from the normal.

The **incident** ray has angle of incidence i.
The **refracted** ray has angle of refraction r.

- When a ray of light enters a glass block it **slows down** and bends **towards the normal**.
- This change in direction is called **refraction**.
- When the ray emerges from the block, it **speeds up** and **bends away from the normal**.
- If the ray meets the surface at 90°, it does not change direction, but there is a change in speed entering and emerging from the glass.

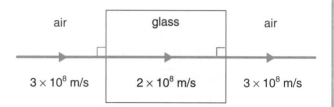

air   glass   air

3×10^8 m/s 2×10^8 m/s 3×10^8 m/s

Lenses

Lenses are **specially shaped pieces of glass or plastic** which are used to refract light in a particular direction.

A converging lens refracts the light so that rays of light are brought together (converge).

CONVEX (**CONVERGING**)

A diverging lens refracts light so that rays of light are made to spread out (diverge).

CONCAVE (**DIVERGING**)

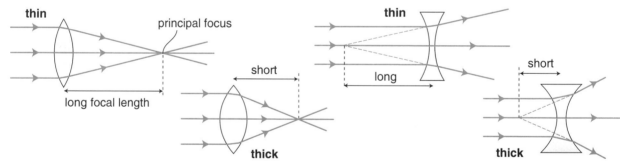

Lens power

Thick lenses bend light **more** than **thin lenses**, and are said to be more powerful. A **powerful** lens has a **shorter** focal length.

$$\text{Power} = \frac{1}{\text{focal length in metres}}$$

$$P = \frac{1}{f} \qquad f = \frac{1}{P}$$

Power is measured in **dioptres (D)**, focal length in **metres (m)**.
Converging (convex) lenses have + power. **Diverging** (concave) lenses have − power.

Top Tip

When calculating the power of a lens, remember to change cm to m.

Measuring focal length

A distant object is considered to give parallel rays. Its image is focussed onto a screen.

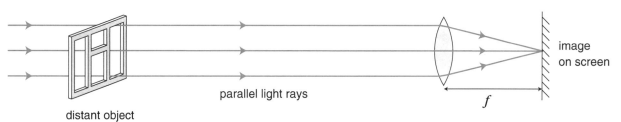

The distance from the spherical convex lens to screen is measured with a ruler.

Ray diagrams

Ray diagrams show how an image is formed with a convex lens.

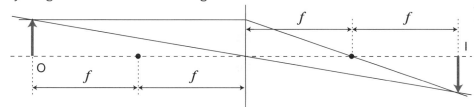

- A ray from the object **O** parallel to the axis bends through the focus point **f**.
- A ray through the centre of the lens goes straight through.
- The image **I** is where the rays meet.

Top Tip
Learn how to measure the focal length of a spherical lens - it is an assessable practical technique!

Image of a distant object:

The rays are only slightly divergent. A **thin** eye lens will focus rays on the retina.

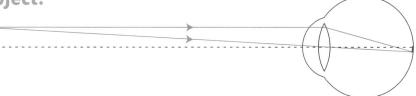

Image of a near object:

The rays are diverging more. The eye lens will need to be **thick** to focus rays on the retina.

The image on the retina is always **upside down** and **laterally inverted**.

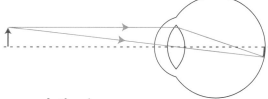

(The ability of our eye to adjust its shape is called **accommodation**.)

Quick Test

1. What is the meaning of refraction?

2. How do the rays bend when going from air to glass?

3. What happens to the speed of light in more dense glass?

4. Name a diverging lens.

5. A lens has a focal length of 20cm. What is its power?

6. What happens to a parallel ray of light going through a convex lens?

Answers 1. change of direction at a boundary **2.** towards the normal **3.** slows down **4.** Concave **5.** 1/ 0.20 = 5D **6.** converges through the focus point

Spectacles to the endoscope

The eye

The **pupil** is an aperture (opening) in the **iris**.

The **cornea** does most of the **refraction** of light.

The **lens** adjusts for near and far vision. It gets **thicker** for **near objects**, and **thinner** for **far objects**.

The **image** is **focussed** on the **retina**.

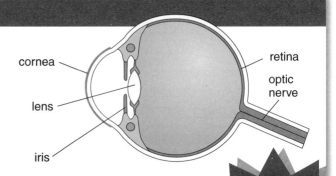

Labels: cornea, lens, iris, retina, optic nerve

Top Tip
Focal length too **long** for **long** sight, too **short** for **short** sight.

Eye defects

The eye can be the wrong shape or the muscles do not adjust the lens enough. Both long and short sight can be corrected using lenses.

Long sight

A **long-sighted person** (usually older) **cannot see close up** clearly. The eye can be too short. The lens can be too thin.

The incoming light is **focussed beyond the retina**. A point becomes a **blur** on the retina. This can be corrected with a **converging** (convex) lens:

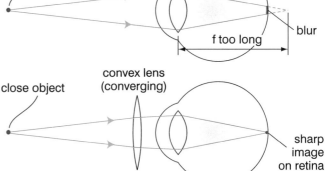

close object ... f too long ... blur

close object ... convex lens (converging) ... sharp image on retina

Short sight

A **short-sighted person** cannot **see far away** clearly. The eye can be too long. The lens can be too fat.

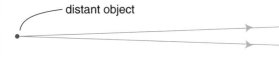

distant object ... f too short ... blur

The incoming light is focussed in front of the retina. A point becomes a **blur** on the retina. This can be corrected with a **diverging** (concave) lens:

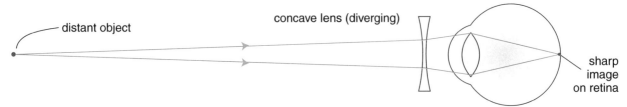

distant object ... concave lens (diverging) ... sharp image on retina

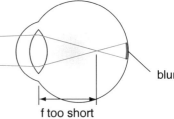

The endoscope

In an endoscope, light is sent from a **light source** down a bundle of fibre optics called the **light guide**.

The light **reflects** inside the patient.

The light travels back up a bundle of **fibre optics** called the **image guide**.

The surgeon can view the image through an **eyepiece** or on a **monitor**.

The light travels by the process of **total internal reflection**. (You met this in telecommunication.)

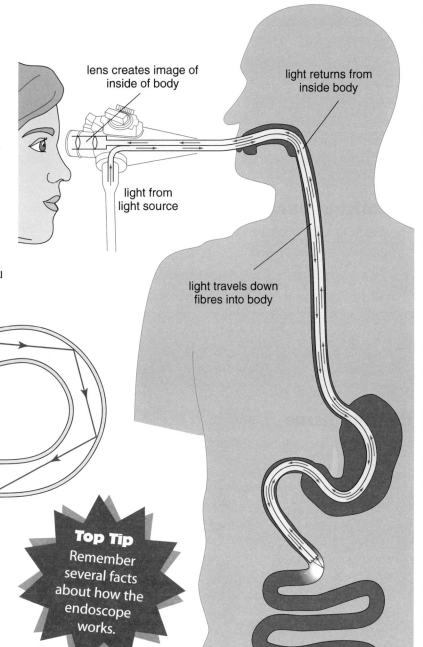

lens creates image of inside of body

light returns from inside body

light from light source

light travels down fibres into body

less dense glass

more dense glass

Fibre optics are **flexible** and can be guided inside the patient.

The light is called a cold light source as no heat is sent down the fibre optics.

Top Tip
Remember several facts about how the endoscope works.

Quick Test

1. What part of the eye accommodates for near and far vision?

2. Where in the eye is the image formed?

3. What fault takes place with the rays of light for a person with long sight?

4. What type of lens would you use in spectacles to correct short sight?

5. How does light travel in an endoscope?

6. What features of an endoscope make it useful?

Answers 1. the lens **2.** retina **3.** rays are focussed behind the retina **4.** diverging **5.** by total internal reflection **6.** flexible, guidable, cold light, small diameter avoids major incision

Using the EM spectrum

Uses of the laser

LASER

The word **Laser** stands for **L**ight **A**mplification by **S**timulated **E**mission of **R**adiation.
A laser produces an **intense** beam of light emitted in one direction.
Different types of laser each produce their own specific wavelength.
The laser finds many uses in medicine.

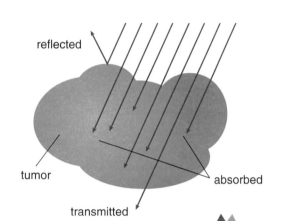

Vaporising tumours

The beam can be **focussed** on a tumour and the tumour will be **vaporised**.

The power and colour (wavelength) of the laser, as well as the type of tissue, have to be considered carefully so that most of the light is absorbed by the tumour, with little reflected or transmitted through. (The neodymium-YAG laser with a wavelength of 1064nm in the infra-red has the right combination of power and wavelength for vaporising tumours.)

reflected

tumor

absorbed

transmitted

Heating tissue

The laser can be used on small pieces of tissue to seal blood vessels by photo-coagulation.

In eye surgery, abnormal blood vessels which grow forward from the retina into the eye can be sealed off. The laser beam is focussed by the eye onto a spot on the retina.

Top Tip
Learn one use of the laser well (e.g. eye surgery) so that you can describe these details.

A surgeon can also use a laser to seal tears and holes in a retina before it becomes detached from the back of the eye.

(An argon laser is used as it emits blue-green light which is absorbed by the red blood).

The laser scalpel

Sometimes very fine surgery is required. Early cancer of the womb and tumours of the voice-box can be treated with a laser beam which offers shallow penetration. (The carbon dioxide laser emits an infra-red beam and can be set to penetrate only 0.1mm of tissue)

Note most types of laser (but not CO_2) can be passed through an endoscope.

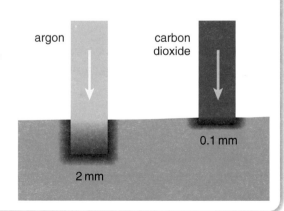

argon

carbon dioxide

0.1 mm

2 mm

Infra-red

Infra-red rays are the **invisible heat rays** given out by all **warm objects**. **Infra-red rays** (or heat waves) are emitted at a **lower frequency** than visible light (**longer wavelength**). On the electro-magnetic spectrum they are found between microwaves and visible light.

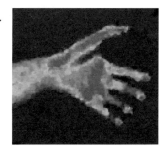

I-R cameras take **thermograms** (colour heat photos) of this radiation. Each colour indicates a different temperature. Malignant tumours are warmer than healthy tissue and this can be seen on the thermogram.

Physiotherapists also use infra-red to heat up muscles and tissue to speed up healing.

Ultra-violet

Ultra-violet (UV) rays are the **invisible rays** given out by **sunlight** or **UV lamps**. UV causes our skin to tan, and too much UV causes **sunburn** and can also lead to **skin cancer**. UV helps us produce vitamin D3 in our skin, can treat skin diseases such as acne and can kill harmful bacteria.
Ultra-violet rays are emitted at a **higher frequency** than visible light (**shorter wavelength**). On the electro-magnetic spectrum they are found between visible light and X-rays.

X-rays and CAT scans

X-rays are emitted at a **higher frequency** than ultra-violet rays (**shorter wavelength**). On the electro-magnetic spectrum they are found between UV and gamma-rays. Overexposure to X-rays can cause cancer. X-rays cannot penetrate lead so radiographers stand behind lead screens or wear lead aprons to prevent overexposure.

Top Tip
Describe to someone the differences between different rays (IR, UV, X-rays). This will help your memory.

X-rays are used for **treatment** or **diagnosis**:
- in treatment, high energy X-rays can be used to damage cancer cells.
- in diagnosis, soft energy X-rays can be used to look at damaged bones inside the body – bones absorb X-rays and the film shows white, and cracks in bone let X-rays through to blacken the film.
- a dentist may also have used X-rays to examine the roots of your teeth.

Computer Axial Tomography (CAT Scan)

X-rays are taken of the body in slices.
The source and detector rotate round the body.
A **3-D picture** is stored on computer and displayed on a monitor.

Quick Test

1. Name 3 uses of a laser.
2. Name 2 uses for infra-red light
3. Give a use and danger for ultra-violet.
4. How can X-rays be detected?
5. What is the main advantage of a CAT scan?

Answers 1. vaporise tumours, heat seal, scalpel **2.** thermograms, speed healing **3.** cure acne, skin cancer **4.** blacken photographic film **5.** 3-D image

Radiation in medicine

Uses of radiation

Diagnosis

Radioactive tracers can be **injected** into the body and followed around the body as they emit rays.

The tracer is **concentrated** in the organ being looked at. It **decays** quickly and **emits γ-rays** which penetrate out of the body. A special gamma camera is positioned over the patient. Signals from the camera build up an image on a monitor screen.

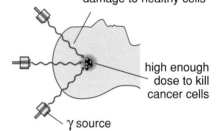

detector

lead blocks

organ

Treatment

Radiation can kill or damage living cells: cancer cells stop reproducing; tumours shrink. The radiation can be high energy X-rays or gamma rays from a radioactive source, e.g. Cobalt 60.

The source rotates around the body. The tumour is hit many times. The healthy tissue has minimal damage.

low dose here causing no damage to healthy cells

high enough dose to kill cancer cells

γ source

Sterilisation

Radiation can be used to kill bacteria or germs. An intense γ-ray source can be used to sterilise boxes of sealed syringes, scalpels or bandages. Gamma rays have the highest frequency (shortest wavelength) on the electro-magnetic spectrum.

Properties of radioactivity

α particles, β particles and γ-rays are all nuclear radiations as they are all emitted from the nucleus.

Type	Nature	Mass (a.m.u.)	Charge	Speed	Absorbed
alpha particle α	2 neutrons + 2 protons	4	+2	$\leq 0.1\,c$	skin, paper or 3cm air
beta particle β	fast electron	$\frac{1}{1840}$	−1	$\leq 0.9\,c$	3cm tissue or 3mm Al.
gamma ray γ	High frequency energy wave	0 (waves have no mass)	0	c	3+ cm lead

(c = velocity of light)

The atom and ionisation

Ionisation

Ionisation occurs when atoms **absorb radiation**. Electrons are removed from the atom to produce **positive** and **negative** ions.

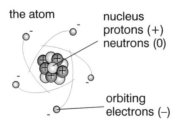

the atom

nucleus
protons (+)
neutrons (0)

orbiting electrons (−)

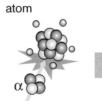

atom

α

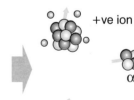

+ve ion

α

electron

The slower, more massive **alpha** causes more ionisation than **beta**. Gamma rays cause the least amount of ionisation and are the most penetrating.

Detection

Most detectors work by ionisation. e.g. **Geiger Muller tube**, **spark counter**, **cloud chamber**. Other detectors include **film badges** and **scintillation counters**.

Activity and half-life

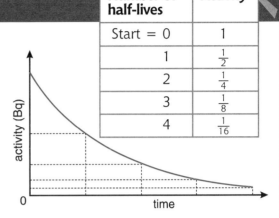

Number of half-lives	Activity
Start = 0	1
1	$\frac{1}{2}$
2	$\frac{1}{4}$
3	$\frac{1}{8}$
4	$\frac{1}{16}$

The activity of a radioactive source is measured in **becquerels** (**Bq**). The **becquerel** is a measure of the **activity** of a radioactive source and is equal to **one disintegration** or **nuclear decay** per second. (There is a **background activity** around us all the time. This activity may be measured on a counter and it has to be allowed for when recording the activity of a source.)

The activity of a radioactive source decays with time.

Half-life

After a certain time a source will have **lost half its activity**. This is called its **half-life**. At each further half-life time the activity drops by half again.

e.g. a substance with an activity of 10 000Bq and a half-life of 10 minutes will drop to 2500Bq in 20 minutes.

Different substances have different half-lives.

Nuclear Medicine	Sodium 24	15 hours
	Iodine 131	8 days
	Cobalt 60	5.3 years
Carbon Dating	Carbon 14	5760 years
Ageing Rocks	Uranium 238	4500 million years

Biological effects

Radiation may cause subtle or severe damage to living tissue. The damage is caused by ionisation. The effect depends on:

- the **total amount of energy** absorbed
- the **nature** of the radiation (e.g. α causes more damage than β or γ) and
- the **type of absorbing matter** (bone, tissue, skin).

The **equivalent dose** takes account of the type and energy of radiation and is measured in **Sieverts** (**Sv**). It is used to compare the biological effect of radiation.

A person should gain protection by **increasing distance**, **shielding**, reducing **exposure time**, and **measurement**.

Quick Test

1. Which type of radiation penetrates the body?
2. How can a tumour receive radiation without it damaging healthy tissue?
3. Which is the most massive radiation?
4. What does radiation cause?
5. What is meant by half-life?
6. What unit measures the biological effect of radioactivity?
7. Name 3 safety measures when dealing with radioactive substances?

Answers 1. γ **2.** source rotates **3.** α **4.** Ionisation **5.** the time for the activity to decay by half **6.** Sieverts **7.** distance, shielding, measurement

Health physics

Test your progress

Use the questions to test your progress.
Check your answers at the back of the book on page 108–109.

1. Sound waves can travel through solids, liquids and gases but not through a ..vacuum

2. a) What is an echo?
   b) If sound waves travel through water at 1500 m/s how deep is the ocean below a ship which hears an echo after just 1s?
   c) What would happen if a shoal of fish swam beneath the ship?

   ....

3. The diagram (right) shows a ray of light travelling through a glass block. Explain what happens to the ray of light as it enters the block. What would happen to the ray if it struck at 0°?

   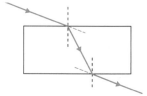

   ....

4. What is ultrasound? Give one use of ultrasound in hospitals.

   ....

5. The diagram (right) shows a ray of light striking the inside surface of a semi-circular glass block.
   a) What is the name of angle A?
   b) What happens to the ray if it strikes the surface at an angle:
      i) smaller than angle A   ii) bigger than angle A?

   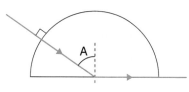

   ....

   ....

6. Explain why a ray of light entering an optical fibre is unable to escape through the glass sides. Give one use for optical fibres.

   ....

7. What are 'ear defenders' and who should use them?

   ....

8. The diagram right shows the penetrating power of three different types of radioactivity. Identify the type of radiation emitted by each of the sources.

   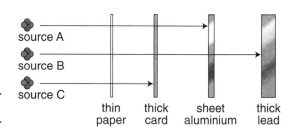

   ....

   ....

9. Which of the three different types of radioactive emission:
   a) is not affected by a magnetic field?   b) carries a negative charge?
   c) creates lots of ions as it travels through matter? d) travels at the speed of light in a vacuum?

   ....

10. Name three medical uses for radioisotopes.

    ....

11. Why is a source of alpha radiation outside the body unlikely to cause damage to vital organs inside the body?

....

12. What range would you expect to find on a clinical thermometer?

....

13. What is the range of hearing of a human?

....

14. What are sounds above the range of human hearing called?

....

15. What instrument could be used to detect sounds from within the human body?

....

16. A person is short-sighted.
   a) Why is the image on the retina a blur?   **b)** What could correct this defect?

....

17. How does the image of the objects we look at appear on our retina?

....

18. A Geiger-Muller tube and counter records 150 counts in 5 minutes of background radiation. Calculate the activity of the radiation.

....

19. What happens to a radioactive source over time?

....

20. A human eye has an effective power of +60D to focus light. What is its effective focal length?

....

21. A lens is placed in front of a light bulb and an image is focussed onto a screen. Why does this not give the focal length of the lens?

....

22. Where would the receiver be found for
   a) An X-ray machine?   **b)** An ultrasound machine?

....

23. What precautions should be taken when working with radioactive sources?

....

24. A radioactive source has an activity of 1200 kBq and a half-life of 30s. What will be the activity after 2 minutes?

....

25. Give two factors on which the effect of radiation absorbed by living materials depends.

....

Input, process and output

Systems and components

Electronics are all around us in everyday systems.

Electronics are also found throughout industry, hospitals and sports.

These all use electronics to monitor, operate or control the functions of everyday life.

Electronic systems are made of smaller components such as resistors, diodes, transistors and capacitors.

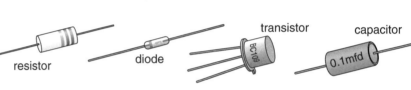

resistor   diode   transistor   capacitor   0.1mfd

These components have been made so small that micro-chips can do the same functions as thousands of these individual components.

microchip

IPO

Electronic systems are divided into three main parts:

IPO = Input Process Output

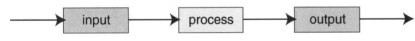

input → process → output

Input = The **start** of the system
Process = **Changes** the signals
Output = Gives us the **result** as we want.

e.g.

System	Input	Process	Output
Hi-fi	CD	Amplifier	Speakers
Electronic thermometer	Temperature sensor	Produce value	LCD display
Radio	Aerial	Radio frequency to audio frequency.	Loudspeaker
Computer	Keyboard	CPU	Monitor
Stopwatch	Switches	Timing circuit	LCD display

Analogue signals

Analogue signals are **continuously variable**.

Most physical quantities such as sound, heat and light are analogue.

A microphone attached to an oscilloscope will display an analogue pattern with speech.

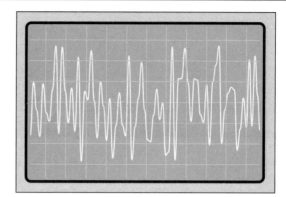

Digital signals

Digital signals have only **2 states**.

These are often called:

ON/OFF, 5V/0V, HIGH/LOW, or **1/0**

A CD player would output a digital pattern on an oscilloscope.

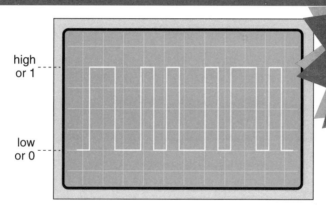

high or 1

low or 0

Top Tip
Digital can have many advantages, but is not always better. Electronics may be digital or analogue.

Conversion

An analogue signal can be converted into a digital signal by **sampling** its amplitude.

The greater the amplitude the greater its binary code.

e.g.

0101 1001 1100 1110
1110 1110 1100 1010
0111 0101 0011 0010
etc.

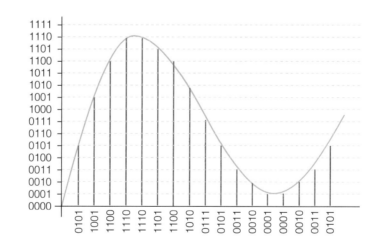

Quick Test

1. Input – _____ – Output. What is the missing word?

2. Name an analogue output device.

3. Name a digital output device.

4. How many states does a digital signal usually have?

5. What continuously changes with sound?

Output devices

Output devices take an electrical signal from the circuit and change it into some useful form of energy, such as light, sound or movement.
→ = 'changes to'

Analogue output devices

The loudspeaker

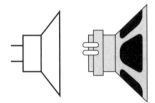

Electrical energy → sound energy

The loudspeaker can handle continuously varying loudness and frequency. It is an **analogue** device.

Loudspeakers are used in hi-fi, radios, TV, telephones and computers.

The electric motor

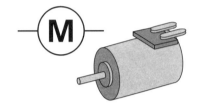

Electrical energy → kinetic (movement) energy

The electric motor can vary its speed. It is an **analogue** device.

Electric motors are used in washing dryers, CD players, printers and windscreen wipers.

The moving coil meter

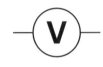

Electrical energy → kinetic energy

The moving coil meter can continuously vary its pointer. It is an **analogue** device.

Moving coil meters are used in ammeters, voltmeters and electric speedometers.

Top Tip
It is estimated there are dozens of electric motors in homes. Find out yours.

Digital output devices

The relay

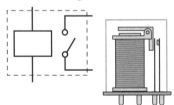

Electrical energy → kinetic energy.

The relay has a coil of wire whose magnetic field operates the opening or closing of a switch. It is a **digital** device.

Relays are used in cars in many places. A low power circuit is used to switch a high power circuit.

The solenoid

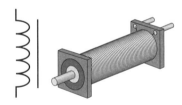

Electrical energy → kinetic energy.

The solenoid provides a straight line movement. An electromagnetic field from a coil pushes a bar out or in. It is a **digital** device.

Solenoids are used in a car for the starter and to operate the locks of a central locking system.

The light emitting diode (LED)

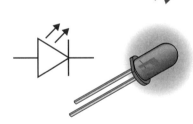

Electrical energy → light energy.

An LED light conducts and is **on** if connected one way or **off** if connected the other way. It is a **digital** device.

LEDs are used in hi-fis, computers, fridges and 7-segment displays.

Using the LED

Light emitting diodes (LED)

LEDs are made from a junction of two **semi-conductor** materials. When electrons cross the junction in the correct direction it emits light. The diode symbol has to 'point' from + to − from a power supply.

Connected the 'wrong' way around = **OFF**.

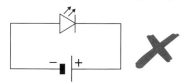 ✗

Connected the 'correct' way around = **ON**.

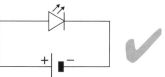 ✔

The LED operates with a small voltage and current, typically 2 V and 10 mA. A resistor normally has to be connected in series with the LED to protect it from high supplies.

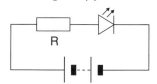

Top Tip
LED is a modern output device. Find out its uses in your home.

The current through the LED and the resistor is the same as they are both in series. The voltage across the resistor = the supply minus the LED voltage.

$$R = \frac{V}{I} = \frac{(5-2)}{0.010} = \frac{3}{0.010} = 300\,\Omega$$

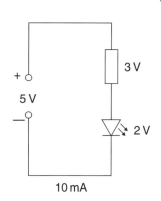

3 V

5 V

2 V

10 mA

The 7-segment display

 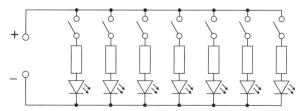

The 7-segment display uses 7 LEDs. Each LED can be operated independently. (Liquid Crystal Displays, LCDs, are also used).

Quick Test

1. What type of energy do output devices convert?

2. Name 2 analogue output devices.

3. Why is a relay a digital device?

4. How does the 'arrow' on the symbol for the LED get connected when it is on?

5. A typical LED is connected to a 6V supply. What value of resistor is required to protect it?

6. What number is formed when all 7 segments of a 7-segment display are on?

Input devices

Converting the world to electrical signals!

Converting to electrical energy

Devices which convert electrical energy directly from sound, heat and light.

The microphone

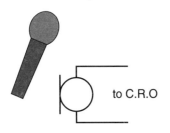

to C.R.O

Sound energy
$\rightarrow$ electrical energy.

The thermocouple

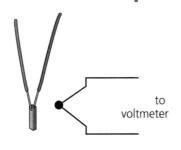

to voltmeter

Heat energy
$\rightarrow$ electrical energy.

The solar cell

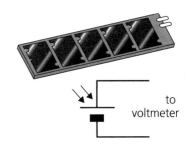

to voltmeter

Light energy
$\rightarrow$ electrical energy.

Controlling electrical energy

Devices where the input alters the voltage of an electrical supply:

The thermistor

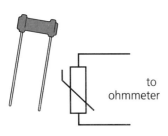

to ohmmeter

As the temperature $\uparrow$,
its resistance $\downarrow$

The light dependant resistor (LDR)

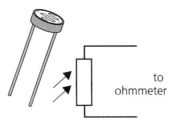

to ohmmeter

As the light $\uparrow$, its resistance $\downarrow$

The switch

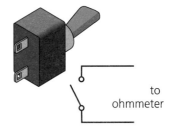

to ohmmeter

Switch open = ∞ resistance.
Switch closed = 0 resistance.

The voltage divider

The voltage from a supply is divided across 2 resistors in the ratio of these resistors.

$$\frac{V_1}{V_2} = \frac{R_1}{R_2}$$

If R_1 or R_2 varies, $V_2 : V_1$ is changed. V will be the input to our **process** section.

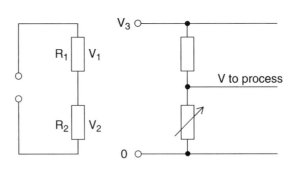

Voltage dividers with an input

Voltage divider with a thermistor

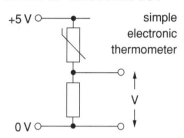

simple electronic thermometer

1. As $T\uparrow$, $R_{therm}\downarrow$
$\Rightarrow V_{therm}\downarrow$, $\Rightarrow V\uparrow$

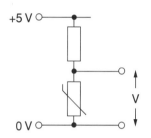

2. As $T\uparrow$, $R_{therm}\downarrow \Rightarrow V\downarrow$

Voltage divider with a LDR

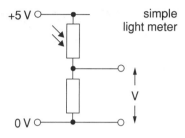

simple light meter

1. As $L\uparrow$, $R_{ldr}\downarrow$
$\Rightarrow V_{ldr}\downarrow$, $\Rightarrow V\uparrow$

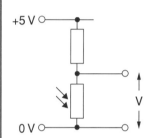

2. As $L\uparrow$, $R_{ldr}\downarrow$, $\Rightarrow V\downarrow$

Voltage divider with a switch

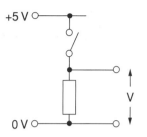

1. S open $\Rightarrow V_S = 5\,V$,
$V = 0\,V$ (low)
S closed $\Rightarrow V_S = 0\,V$,
$V = 5\,V$ (high)

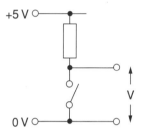

2. S open $\Rightarrow V = 5\,V$ (high)
S closed $\Rightarrow V = 0\,V$ (low)

Voltage divider with a capacitor

A capacitor stores electric charges on its plates.
As the charge increases, the voltage increases.
A capacitor takes **time** to charge up.
The capacitor acts as a **time delay** input:

The **time** to charge **increases** when:
1. C increases (a bigger capacitor takes longer to fill / charge).
2. R increases (a bigger resistor $\Rightarrow$ a smaller charging current).

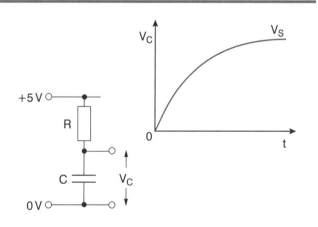

Quick Test

1. What is the energy change in a solar cell?

2. What happens to the resistance of a LDR as the light decreases?

3. What is the symbol for a thermistor?

4. In a voltage divider what ratio does the voltage ratio copy.

5. What voltage does the voltage across a capacitor eventually reach?

6. A 12V supply is divided by a 5Ω and a 1Ω resistor in series. What is the voltage across the 1Ω?

Answers 1. Light energy to electrical energy 2. Increases 3. See above 4. R1:R2 5. supply voltage 6. 2V

55

Electronics

Sensor with

circuits

an electronic switch

The ... tage controlled switch respona... ige at the input.

As C is moved ... n B to A, V increases from 0 V to 5 V.

When $V_{be} > 0.7$ V then the transistor **conducts** from c to e and the **LED has been switched on**.

When $V_{be} < 0.7$ V or negative then the transistor **does not conduct** from c to e and the LED goes **off**.

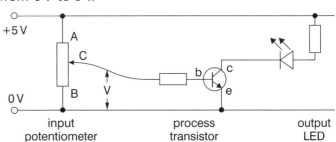

input potentiometer   process transistor   output LED

Light controlled circuits

Cover the LDR:

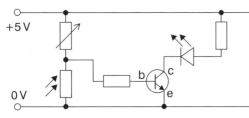

as L $\downarrow$, $R_{ldr}\uparrow$, $V_{ldr}\uparrow$ >0.7 V $\Rightarrow$ transistor switches to conduct $\Rightarrow$ LED goes **on**.

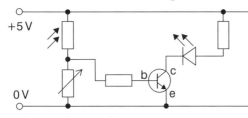

as L $\downarrow$, $R_{ldr}\uparrow$, $V_{ldr}\uparrow$, V_R <0.7 V $\Rightarrow$ transistor does not conduct $\Rightarrow$ LED goes **off**.

What happens in each circuit when light increases? A light increase could switch on a security warning light. A light decrease could switch on a night light. The variable resistor is used to set the light level at which the transistor switches **on** or **off**.

Switch controlled circuits

Use a resistor and switch as inputs when you design and explain circuits like those above.

Temperature controlled circuits

Warm the thermistor:

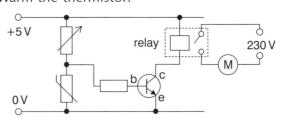

as T $\uparrow$, $R_{therm}\downarrow$, $V_{therm}\downarrow$ <0.7 V $\Rightarrow$ transistor does not conduct $\Rightarrow$ relay goes **off**.

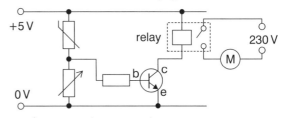

as T $\uparrow$, $R_{therm}\downarrow$, $V_{therm}\downarrow$, V_R >0.7 V $\Rightarrow$ transistor conducts $\Rightarrow$ relay goes **on**.

What happens in each circuit when temperature falls? A temperature increase could switch on a fan. A temperature decrease could switch on a heater.

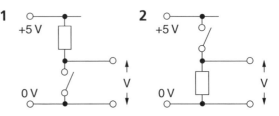

Time controlled circuits

Connect a wire across the capacitor to discharge it. Remove the wire to start it charging.

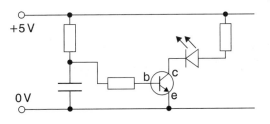

As the capacitor charges, the voltage across it increases. Only when the voltage reaches a certain value (0.7 V) does the transistor conduct (switch on) and the LED goes on. This could be used for a delay for a pedestrian crossing or altered to delay a car's interior lights going off.

Logic gates

Top Tip
The keypad on a school drinks machine will use logic gates. Can you see how?

Logic gates are made from transistors and resistors on tiny Integrated Circuits (chips <1mm²).

Connecting an input to 5V = logic **1**   Connecting an input to 0 V = logic **0**.

NOT gate or INVERTER

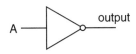

A	Output
0	1
1	0

The output is **NOT** the input.

AND gate

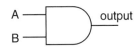

A	B	Output
0	0	0
0	1	0
1	0	0
1	1	1

The output is high when A is high **AND** B is high.

OR gate

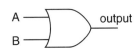

A	B	Output
0	0	0
0	1	1
1	0	1
1	1	1

The output is high when A is high **OR** B is high **OR** both high.

Gates make many decisions: A car ignition switch **AND** door closed switch ⇒ car will start.

Combinational logic

A light has to come on automatically when it's dark if its system has been switched on. The switch gives a '1' when on and the light sensor gives a '1' in daylight. Combine an AND gate with a NOT gate.

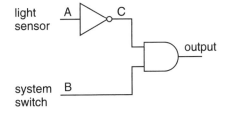

A	B	C	Output
0	0	1	0
0	1	1	1
1	0	0	0
1	1	0	0

Quick Test

1. What is the main purpose of a transistor in electronics?

2. Why is a LDR in series with a resistor at the input stage?

3. What is the purpose of the variable resistor?

4. How do you make a time delay longer?

5. What is a truth table?

Answers 1. automatic switch **2.** To divide the supply voltage **3.** To control the physical level (e.g. how dark) for switching **4.** Increase R or C **5.** A table of inputs and outputs

Pulses and counting

The clock pulse generator

An **oscillator** is a **digital device** which produces a series of **clock pulses**. Clock pulses are used in digital watches and computers.

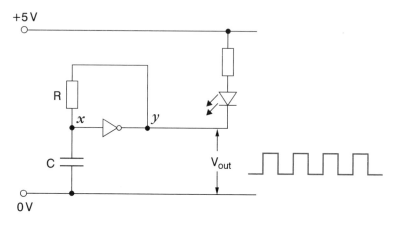

How it works

1. At the start, the capacitor is uncharged, logic at $x = 0$. Output y from the inverter $= 1$ (5 V), both sides of LED $= 1$, **no voltage** across it, the LED is **off**.

2. The output from the inverter pushes charge through the resistor to the capacitor. The capacitor charges till $x \rightarrow 1$ and $y \rightarrow 0$. The LED **conducts** and is **on**.

3. As $x = 1$ and $y = 0$, the capacitor now discharges through the resistor till $x \rightarrow 0$ and $y \rightarrow 1$. The LED goes **off**.

4. The capacitor starts to charge again... When $x \rightarrow 1$ and $y \rightarrow 0$, the LED goes **on** again.

5. Steps 3 and 4 repeat producing pulses.

Counting the pulses

A **counter** is the electronic circuit to count the pulses. The counter takes in the clock pulses and sends out a 4-bit binary pattern.

No. of Pulses	D	C	B	A
0	0	0	0	0
1	0	0	0	1
2	0	0	1	0
3	0	0	1	1
4	0	1	0	0
5	0	1	0	1
6	0	1	1	0
7	0	1	1	1
8	1	0	0	0
9	1	0	0	1

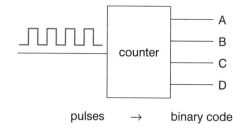

pulses → binary code

Top Tip
Remember Signal order: Pulses – Binary – Decimal

The decoder and the 7-segment display

Top Tip
Device Order:
Clock Pulse
Generator – Counter –
Decoder –
7-segment
Display

Binary code can be hard to understand.

A **decoder** will take the **4-bit binary** code and change it to switch the **correct inputs** of a 7-segment display for **decimal** numbers.

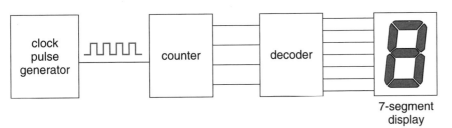

7-segment display

Binary to decimal

Decimal numbers: e.g. 2345
$$= (2 \times 10^3) + (3 \times 10^2) + (4 \times 10^1) + (5 \times 10^0)$$
$$= (2 \times 1000) + (3 \times 100) + (4 \times 10) + 5$$
$$= 2000 + 300 + 40 + 5$$

Binary numbers: e.g. 1111
$$= (1 \times 2^3) + (1 \times 2^2) + (1 \times 2^1) + (1 \times 2^0)$$
$$= (1 \times 8) + (1 \times 4) + (1 \times 2) + (1 \times 1)$$
$$= 8 + 4 + 2 + 1 = 15$$

Can you see how to change the following binary numbers to their decimal equivalent?

Binary Code				Decimal Number
$2^3 = 8$	$2^2 = 4$	$2^1 = 2$	$2^0 = 1$	
0	0	0	0	0
0	0	0	1	1
0	0	1	0	2
0	0	1	1	3
0	1	0	0	4
0	1	0	1	5
0	1	1	0	6
0	1	1	1	7
1	0	0	0	8
1	0	0	1	9

Quick Test

1. What devices could use a clock pulse generator?
2. If there are pulses every 1/6 s, what is their frequency?
3. What is the range of numbers in decimal?
4. What is the binary of decimal 2?
5. What is the range of numbers in binary?
6. What is the decimal of 0101?
7. What device makes the 7-segment display only require 4 switches?

Answers 1. Digital watch and computer 2. 6Hz 3. 0–9 4. 0010 5. 0–1 6. 5 7. Decoder

Analogue processes

Amplifiers are usually analogue devices and their process is to amplify an **electrical** signal.

Amplifiers

Amplifiers play an important part in many devices.

Amplifier gain

The purpose of an amplifier is to increase the **amplitude** of an **electrical** signal. There should be **no change** in **frequency** or **pattern** or the signal is said to have distortion.

An amplifier works with a power supply. The power supply provides the energy for the bigger signal. Basically the pattern of the input signal is transferred to the larger electrical supply signal.

A series of transistors is built into an integrated circuit in modern amplifiers.

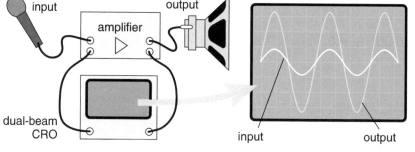

Top Tip
Whether Voltage or Power,
$$Gain = \frac{Output}{Input}$$

Voltage gain

The **voltage input** and the **voltage output** can be compared using oscilloscopes or multi-meters.

The **voltage gain** of an amplifier is then found from:

$$Voltage\ Gain = \frac{Voltage\ Output}{Voltage\ Input}$$

As gain is a **ratio**, there are **no units** for gain.

A weak voltage signal in an electronic system can be increased, e.g. TV aerial booster.

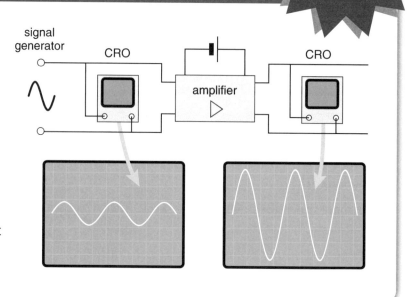

Power gain

In a PA (public address) system or hi-fi, it is the **power** to drive the speakers that is important.

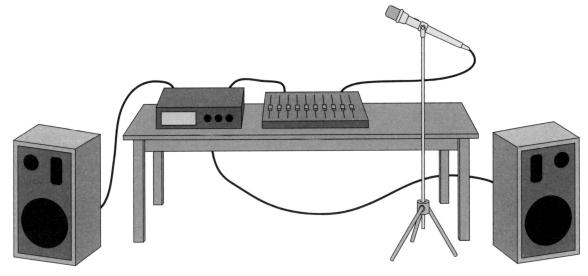

A mixer or pre-amp will make sure all the input signals are balanced ready for the **power amplifier**.

The **power gain** of an amplifier is then found from:

$$\text{Power Gain} = \frac{\text{Power Output}}{\text{Power Input}}$$

Amplifier circuits have resistance and the coil of a loudspeaker also has resistance.

To calculate the **power input** or the **power output** the most useful equation is:

$$P = \frac{V^2}{R}$$

The **input resistance** of an amplifier is usually quite **high** – watch out for kΩ.

The **resistance** of a loudspeaker is often around **8Ω**.

The **input signal** to an amplifier is often measured in **mV**.

Quick Test

1. What type of signal does an amplifier increase?
2. What happens to the frequency of the signal during amplification?
3. What part of a signal is amplified in an amplifier?
4. The voltage input to an amplifier is 10 mV and the output voltage is 1.5V. What is the gain?
5. What power of amplifier has supplied 10 V to a 4 Ω loudspeaker?

Answers 1. Electrical 2. No change 3. Amplitude 4. 150 5. 25 W

Test your progress

Use the questions to test your progress.
Check your answers at the back of the book on page 108–109.

1. What are the three main parts of an electronic system called?

....

2. What particles move through electronic circuits?

....

3. Name three input devices.

....

4. Name three output devices.

....

5. What do the two arrows leaving a diode symbol tell us?

....

6. Draw the circuit symbol for a transistor.

....

7. Name two types of electronics.

....

8. An amplifier takes in a signal whose voltage is 0.02V. If the output voltage is 1.8V what is the voltage gain of the amplifier?

....

9. What type of signal is said to be continuously variable?

....

10. What type of signal is said to have only two states?

....

11. An LED appears to be connected to a 1.5V cell without any breaks in the circuit. If it is not lit suggest the most likely cause.

....

12. If an LED has to be connected to a 6V battery, what should be connected in series with it?

....

13. Which logic gate should be used in an alarm which should go off when the front door or the back door are opened?

....

14. Which logic gate should be used in a garden hedge trimmer which requires the left and right hands to hold two switches on?

....

15. Draw the circuit symbol for an inverter.

....

16. Draw the truth table for an AND gate.

17. What is a light-dependent resistor? Name one use for a light-dependent resistor.

....

18. A light beam operates an automatic circuit. What component will the beam of light fall on and how will it change?

....

19. What is the function of a transistor in modern electronics?

....

20. How does a transistor operate?

....

21. What value of resistor is required to be in series with an LED when it requires 2V and draws 15mA from a 5V supply?

....

22. What components are required to construct a pulse generator?

....

23. How can the frequency of a pulse generator be increased?

....

24. What type of output is sent from a counter circuit?

....

25. Why is a 7-segment display called this?

....

CREDIT

Speed

Speed and velocity

We often give a direction when we use the term velocity, but here we will usually use the word velocity as if it were the same as speed.

I travel 8 m every second. My speed is 8 m/s.

This train is travelling north at a speed of 40 m/s. Its velocity is 40 m/s north.

Key Fact

1. $v = \dfrac{d}{t}$

2. Velocity is the speed of an object in a particular direction.

How to calculate speed

The **speed** (or **velocity**) of an object is defined as the distance travelled in 1 second.

To find the speed of an object we need to know how far it has travelled and how long it took to travel this distance. Then we use the equation:

$$\text{speed} = \frac{\text{distance}}{\text{time}} \text{ or } v = \frac{d}{t}$$

We can write this equation as a formula triangle. We cover the quantity the question is asking us to calculate. The triangle now shows us the formula we should use.

Example

A sprinter runs 400 m in 50 s. Calculate his speed.

$v = \dfrac{d}{t}$

$v = \dfrac{400 \text{ m}}{50 \text{ s}}$

$v = 8$ m/s

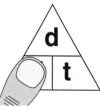

Example

A cricket ball travels at 50 m/s for 2 s after being hit. How far has the ball travelled?

Using the formula triangle we see that **distance = speed × time**, or $d = v \times t$

$d = v \times t$

$d = 50 \times 2$

$d = 100$ m

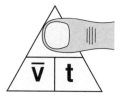

Top Tip
Practise using your formula triangle. It is very useful for many formulae you will need in your exams.

Example

A car travels 200 km at an average speed of 40 km/h. How long does the journey take?

Using the formula triangle we see that

$$\text{time} = \frac{\text{distance}}{\text{speed}} \text{ or } t = \frac{d}{v}$$

$t = \dfrac{d}{v}$

$t = \dfrac{200}{40}$

$t = 5$ hr

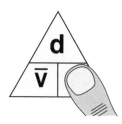

Measuring average speed

Average speed $\bar{v}$ is the total distance travelled over the total time taken.

($\bar{v}$ is pronounced v bar)

$$\bar{v} = \frac{d}{t}$$

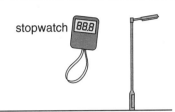

stopwatch

tape

e.g. to measure the average speed of a cyclist on a road we would use a measuring tape and a stop-clock:

- Use the tape to measure a **marked distance**.

- Use the stop-clock to measure the **time taken**.

- Then we use the formula $\bar{v} = \frac{d}{t}$ to calculate the **average speed**.

Top Tip

Try to think of situations where average and instantaneous speeds are different.

Measuring instantaneous speed – using a light-gate

Instantaneous speed is the speed at a certain time.

A good estimate of instantaneous speed is obtained by using a **very small time interval**.

To measure the speed of a toy car a light-gate (photocell and light beam) is attached to an electronic timer or computer timer.

A card is attached to the toy car. The length of card passes through the light beam.

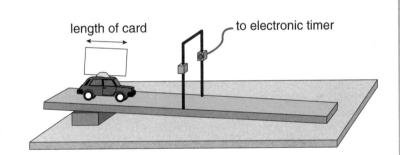

length of card

to electronic timer

- Use a ruler to measure the **length of card**.

- Use the light gate and electronic timer to measure the **short time taken**.

- Then we use the formula $v = \frac{d}{t}$ to calculate the **instantaneous speed**.

Quick Test

1. What two measurements do you need to calculate the speed of an object?

2. Name 2 units you could use to measure the speed of an object.

3. Calculate the speed of a man who runs 70 m in 14 s.

4. How long will it take a man running at 10 m/s to travel 550 m?

5. How far will a car travel in 3h if its speed is 60 km/h?

Answers 1. Distance travelled and time taken 2. m/s, km/h 3. 5 m/s 4. 55 s 5. 180 km

Acceleration

What is acceleration?

Acceleration is defined as the **rate of change of speed**.
Acceleration is the change of speed in unit time (1 s).

If an object is **changing** its speed or velocity it is **accelerating**. The acceleration of an object tells us **how rapidly its speed is changing**.

My speed is increasing by 10 m/s each second. My acceleration is 10 m/s².

Examples

A motorcyclist has an acceleration of 20 km/hr per s.
This means his speed increases by 20 km/h each second.
A rocket has an acceleration of 50 m/s per s (sometimes written as 50 m/s²).
This means that the rocket increases its speed by 50 m/s every second.

To calculate the acceleration of an object we need to know by how much its speed or velocity has changed and how long this change in velocity has taken. Then we use the equation:

$$\text{acceleration} = \frac{\text{change in velocity}}{\text{time taken}} \quad \text{or} \quad a = \frac{\Delta v}{t} \quad \text{or} \quad a = \frac{v - u}{t}$$

where the Δ symbol (the Greek letter 'delta') means 'change in' and where v is the final velocity and u is the starting velocity.

Top Tip
The average velocity during a steady acceleration can be found using $\bar{v} = \frac{(u + v)}{2}$

Example

A racing car accelerates from rest to a speed of 100 m/s in just 5 s. Calculate the acceleration of the car.

$$a = \frac{\Delta v}{t}$$
$$a = \frac{100}{5} = 20 \text{ m/s}^2$$

Example

The bobsleigh team is about to increase its speed from 0 m/s to 50 m/s in 10 s. Calculate its acceleration.

$$a = \frac{\Delta v}{t}$$
$$a = \frac{50}{10} = 5 \text{ m/s}^2$$

Example

A car accelerates at 4 m/s² for 8 s. If its initial speed is 6 m/s calculate its final speed.

$$a = \frac{\Delta v}{t} = \frac{v - u}{t}$$
$$4 = \frac{v - 6}{8}$$
$$32 = v - 6$$
$$v = 38 \text{ m/s}$$

Measuring average acceleration

Using 2 light-gates and a stop-clock.

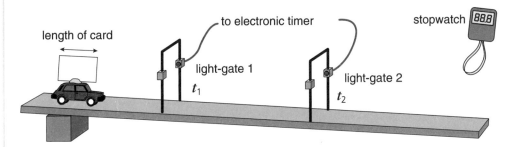

A short length of card is attached to the vehicle to cut the light beam.
- At light-gate 1 we use the **length of card** and **time** to obtain an **initial velocity u**.
- At light-gate 2 we use the **length of card** and **time** to obtain a **final velocity v**.
- The stop-clock is used to **record the time t between velocities**.
- Then we can use the equation $a = \dfrac{v - u}{t}$ to calculate the average acceleration between the light-gates.

Measuring acceleration at a point

Here we use a double card with single light-gate attached to a motion computer.

The second card cuts the light beam quicker than the first.

The length of card needs to be measured and entered to the motion computer.

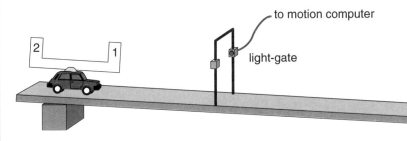

The motion computer records:
- the **time** for the **first card**
- the **time** for the **second card**
- the **time between the cards**.

It can then calculate the acceleration. Can you see how?

Top Tip
The motion computer needs to know the velocity and acceleration equations.

Quick Test

1. Calculate the acceleration of a Fiat which increases its speed by 60 m/s in 20 s.

2. A VW is travelling at 10 m/s when it accelerates at 5 m/s² for 3 s. What is its new speed?

3. A bus is travelling at 20 m/s when it decelerates at 2 m/s². How long does it take to stop?

Graphs of motion

We can use a speed- or velocity-time graph to show the journey of an object.

Speed-time graphs

Describing motion from a speed-time graph

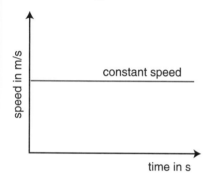
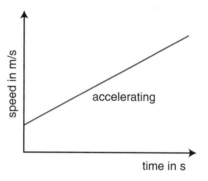
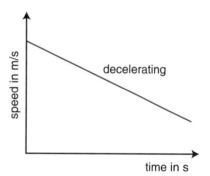

Horizontal line: object moving at **constant speed**.

Straight line sloping upwards: object increasing speed: **constant acceleration**.

Straight line sloping downwards: object decreasing speed: **constant deceleration**.

Calculating distance from a speed-time graph

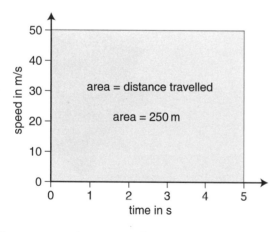
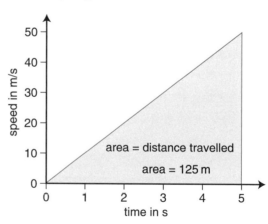

The **area** under a speed-time graph shows the **distance** an object has travelled.

$$d = \text{area under a } v/t \text{ graph}$$

Calculating acceleration from a speed-time graph

We can find the **acceleration** of an object by measuring the **gradient** of its velocity-time (or speed-time graph).

a = gradient of a v/t graph.

The acceleration of the object in the graph on the right is:

$$\frac{30\,\text{m/s}}{5\,\text{s}} = 6\ \text{m/s}^2$$

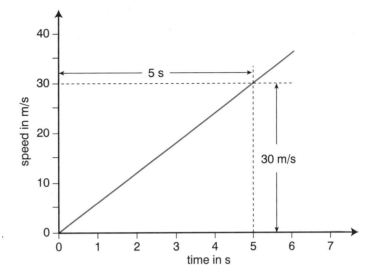

Example

A motorcyclist starting from rest accelerates to a speed of 40 m/s in 4 s. He travels at this speed for 10 s before decelerating to a halt in 8 s.

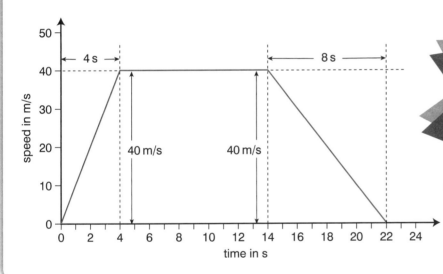

Top Tip

If you have to draw a graph remember to:
• use a sharp pencil and don't press too hard. You may want to rub it out!
• use a ruler for straight lines and axes.
• label the axes and include units.

Quick Test

1. On a speed-time graph what do the following show:
   a) a horizontal line?
   b) a straight line sloping steeply upwards?
   c) a straight line sloping gently downwards?

2. Draw a speed-time graph to describe the following journey:
   • A sprinter starting from rest accelerates to a speed of 10 m/s in 2 s.
   • He travels at this speed for the next 8 s then decelerates to 2 m/s in 4 s.
   • He continues to jog at this speed for the next 6 s.

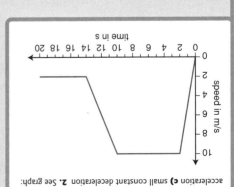

Answers 1. a) constant speed **b)** large constant acceleration **c)** small constant deceleration **2.** See graph.

Forces

Effects of forces

A force, applied to an object, has the ability to change the

• **shape** • **speed** • **direction of movement**

of an object.

Recognising forces

pushing off the blocks   **pulling** the grass roller   **compressing** the spring

twisting the bottle   **lifting** the bag   **stretching** the catapult

Measuring forces

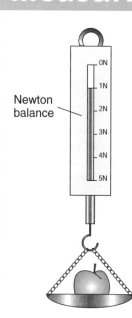

Newton balance

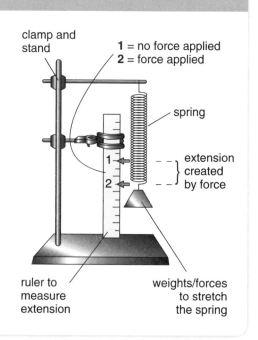

clamp and stand

1 = no force applied
2 = force applied

spring

extension created by force

ruler to measure extension

weights/forces to stretch the spring

If we apply a force to a spring or wire it will extend or stretch. The extension varies with the applied force.

The extension is the increase in length of the spring/wire when a force is applied to it.

We can measure the size of a force using a **Newton balance**. This consists of a spring and a scale; the scale measures how much the spring stretches when a force is applied to it.

The larger the force the more the spring stretches. We measure the forces in **newtons** (N). An average sized apple weighs about 1N.

Mass, gravity and weight

Mass

Mass is **the amount of matter** there is in an object.
Mass depends on the number and type of atoms.
Mass is a **scalar** – it has only size, no direction.
We measure mass in **kilograms** (**kg**).

Lifting mass

On Earth, a **force of 10 N** is required to lift
every **1 kg of mass**.

Gravitational field strength

On Earth, **every kg of mass** has a weight of **10 N**.
The weight per unit mass is called
the **gravitational field strength**.

$$g = \frac{W}{m}$$

On Earth $\boxed{g = 10\,N/kg}$ On the Moon $\boxed{g = 1.6\,N/kg}$

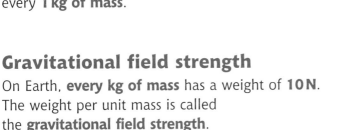

Top Tip
Learn when to say mass and when to say weight.

Weight

Weight is a **force**.
Weight is the Earth's pull on an object.
Weight acts downwards.
Weight depends on mass and the gravitational field strength.

$$W = m \times g$$

m is the mass of the object
g is the strength of gravity

On Earth, $g = 10\,N/kg$
(i.e. an object with a mass of 1 kg
weighs 10 N on Earth).

On the Moon, $g = 1.6\,N/kg$
(i.e. an object with a mass of 1 kg
weighs 1.6 N on the Moon).

Example

This block has a mass of 6 kg. It
contains 6 kg worth of matter.
On Earth its mass is 6 kg and its weight is:

$W = m \times g = 6 \times 10 = 60\,N$

On the Moon the amount of matter in the block,
i.e. its mass, does not change. It is still 6 kg. But
its weight is less:

$W = m \times g = 6 \times 1.6 = 9.6\,N$

Quick Test

1. What does a spring do to measure force?
2. What three things can a force change?
3. What is the main unit of mass?
4. What is the weight, on Earth, of
   a) 100 g? b) 500 g? c) 1 kg? d) 10 kg?
5. Why do you think does the Moon not have as strong a pull on mass as the Earth has?

Answers 1. Extend 2. Speed, shape, distance 3. kg 4. a) 1 N b) 5 N c) 10 N d) 100 N 5. The Moon has less mass.

The force of friction

Friction

Friction opposes the motion of a body.
Whenever an object moves or tries to move, **friction** is present.

All moving objects cause friction to occur.

friction

Friction between surfaces can make them hot and wear them away.

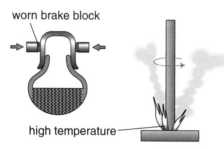

worn brake block

high temperature

Friction between the tyres of a car and the surface of a road is very important. If there is insufficient grip it is impossible to stop or steer the car safely.

Streamlining and lubricating

- As a bobsleigh travels down the run it gains speed.
- There are large frictional forces at work between the sleigh and the air, and between the runners and the ice.
- To keep these forces to a minimum the bobsleigh is:
  a) streamlined. It is shaped so it cuts through the air with less resistance.
  b) the runners are coated with a **lubricant**, such as wax.

Moving through air

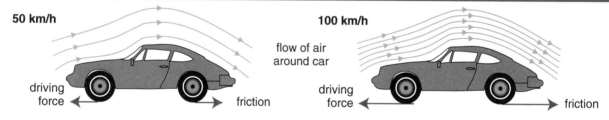

50 km/h

driving force   friction

flow of air around car

100 km/h

driving force   friction

- When an object moves through air or water it will experience **frictional** or **resistive forces** (drag) which will try to prevent its motion.
- The faster the object moves the larger these resistive forces become.

Stopping a vehicle

In order to slow or stop a vehicle a braking force needs to be applied to it. This is usually achieved by using the friction between surfaces.

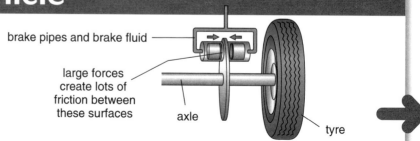

brake pipes and brake fluid

large forces create lots of friction between these surfaces

axle

tyre

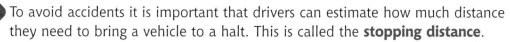

To avoid accidents it is important that drivers can estimate how much distance they need to bring a vehicle to a halt. This is called the **stopping distance**.

The stopping distance consists of two parts:

1. The **thinking distance**: this is the distance a vehicle travels before a driver applies the brakes.
2. The **braking distance**: this is the distance the vehicle travels whilst braking.

> Stopping distance = thinking distance + braking distance.

Thinking distance

Things that will affect the thinking distance are:

- The **speed of the vehicle**: the greater the speed the greater the distance travelled before the brakes are applied.
- The **reaction time of the driver**: the slower a driver's reactions the greater the distance travelled before the brakes are applied. Drinking alcohol, taking drugs and tiredness will increase this time.

Braking distance

Things that will affect the braking distance are:

- The **mass** of the vehicle. The greater the mass, the greater the distance needed to stop.
- The **speed** of the vehicle. If a driver doubles his speed, he will need at least four times more braking distance.
- The **braking force** applied to the wheels. Faulty brakes can result in smaller forces being applied to the wheels. But if too much force is applied, the wheels can skid.
- **Frictional forces** between the tyres and the road. If these forces are reduced, the braking distance will increase. This can happen in bad weather conditions (wet or icy), if the tyres are worn, or if the road surface is smooth.

Stopping distance

Speed	Thinking distance	Braking distance	Total stopping distance
At 13 m/s (30 mph)	9 m	14 m	23 m
At 22 m/s) (50 mph)	15 m	38 m	53 m
At 30 m/s (70 mph)	21 m	75 m	96 m

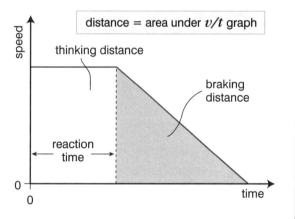

distance = area under v/t graph

Quick Test

1. What is friction?
2. In which direction does friction act?
3. What is streamlining?
4. Name two possible effects of friction between two surfaces.
5. Which two distances are added together to find the stopping distance of a car?
6. Name two things that may affect the thinking distance of a driver.
7. Name three things that may affect the braking distance of a car.

Answers 1. A force that opposes motion 2. The opposite direction to motion 3. Shaping to reduce resistance 4. Heat and wearing away of the surface 5. Thinking and braking distance 6. Speed of vehicle, reaction time 7. Mass of vehicle, speed, weather conditions

Newton's laws

Balanced forces

If several forces are applied to an object, they may **cancel each other out**. The forces are **balanced**.

- If the forces applied to an object are **balanced** they will have **no effect on its motion**.
- If the object is stationary it will **remain stationary**.

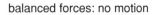

balanced forces: no motion

- If the object is moving it will **continue to move in the same direction** and at the **same** speed.

Example
If the driving force of this aircraft equals the drag, it will travel at a constant speed. If the lift force equals the weight, the aircraft will stay at a constant height.

stationary objects

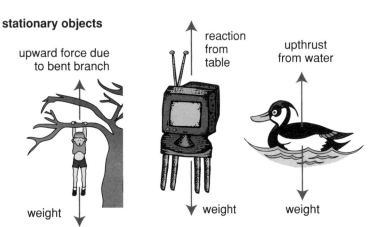

upward force due to bent branch

reaction from table

upthrust from water

weight   weight   weight

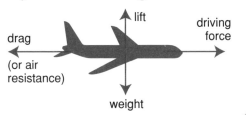

lift

drag

driving force

(or air resistance)

weight

Newton's first law of motion – NI

An object will **remain at rest** or will **remain at constant speed in a straight line** unless acted on by an **unbalanced force**.

NI is a **no force** law.
Balanced forces are equivalent to no force at all. The speed remains the same.

In space objects keep going without energy or rockets.
A force is only used when we want a **change**.

Normally we expect objects to stop. Newton says something must be doing the stopping. This can be the **friction** between the object and the ground.

Seat belts

Newton I says an object should keep its speed. In a crash, if we are standing on a bus, or do not have a seat belt on, NI says we **will just keep going** – if the vehicle stops. We are **not** 'thrown forward'.

A seat belt applies a force in the opposite direction of motion which decelerates the person with the vehicle. A seat belt will have some 'give' so as not to cause injury.

Unbalanced forces

Forces applied to an object which do not cancel out are **unbalanced**.
Unbalanced forces cause **acceleration**.

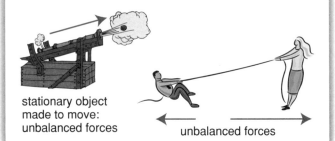

stationary object made to move: unbalanced forces

unbalanced forces

Forces and acceleration

An object whose motion (speed or direction) is changing is accelerating.

The larger the force the greater the acceleration – **direct proportion**.

The larger the mass the smaller the acceleration – **inverse proportion**.

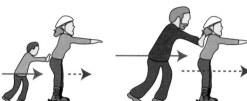

small force → smaller acceleration

large force → larger acceleration

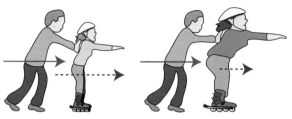

small mass → larger acceleration

large mass → smaller acceleration

Newton's second law of motion – NII

The **acceleration** of an object varies **directly** with the **unbalanced force** and **inversely** with its **mass**.

$$a = \frac{F}{m} \qquad F_{un} = ma \qquad \text{NII}$$

Top Tip

Balanced forces are equivalent to no force and cause no change in speed or direction. Unbalanced forces cause change.

Terminal velocity

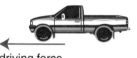

driving force

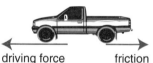

driving force   friction

driving force   friction

The driver presses the accelerator. The driving force makes the car accelerate.

The accelerator is kept down. Air resistance increases ⟹ a smaller acceleration.

The accelerator is kept down. Air resistance **balances** driving force.

The final, constant speed of the car is known as **terminal velocity**.

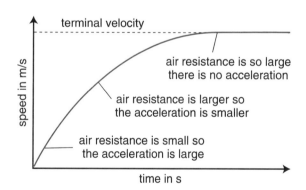

terminal velocity

air resistance is so large there is no acceleration

air resistance is larger so the acceleration is smaller

air resistance is small so the acceleration is large

speed in m/s

time in s

Quick Test

1. What effect do balanced forces have on the motion of an object?

2. What does the size of the acceleration of an object depend on?

3. Why does a car have a top speed?

4. If a 2 kg block is pulled with a force of 10 N and friction is 2N,
   a) what is the size of the unbalanced force?
   b) what is the block's acceleration?

Answers 1. No overall effect **2.** Mass of object, size of force **3.** Air resistance balances engine force **4. a)** 8N **b)** 4m/s²

Work, energy and power

Energy transformations

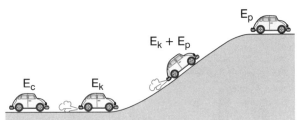

- The petrol has **chemical potential energy**.
- The **kinetic energy** of the car is increasing.
- Work is being done against friction – the surroundings are **gaining heat**.
- The car now has **gravitational potential energy**.

Energy is never created or destroyed – just changed from one type to another. This is called **conservation of energy**.

Work done

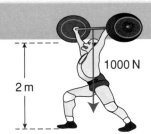

Work is done when **energy is transferred**. Both **work** and **energy** are measured in **Joules (J)**.

To do work: apply a **force** through a **distance**. The object gains energy.

Work done = Force × Distance moved.

$$E_w = F \times d$$

The weightlifter has done work.

$$W = F \times d$$
$$= 1000 \times 2 = 2000 \, J.$$

It is useful to write the equation as a formula triangle:

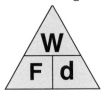

Kinetic energy

Kinetic energy is the energy of moving objects.

A vehicle is accelerating. **Work** is being done. The vehicle gains **kinetic energy**.

$$E_k = \tfrac{1}{2}mv^2$$

The energy depends on **mass**:

- A car and a lorry are going at the same speed. The lorry has more mass. It will take more work to stop. The **more massive lorry** has **more kinetic energy**.

The energy depends on the **square of the velocity**:

- Two identical cars. One is going twice as fast as the other. $2^2 = 4$. It will have **four times the energy**. It will take four times as much work to stop.

How much more energy will a vehicle need to become three times as fast? ... $3^2 = 9$ times.

Example

A 1000 kg car is going at 30 m/s (70 mph). How much energy will be needed to stop it?

$$E_k = \tfrac{1}{2}mv^2 = \tfrac{1}{2}1000 \,(30^2) = 500 \times 900 = 450\,000 \, J.$$

This energy will be **lost as heat to the surroundings**.

Top Tip
Remember to only square the velocity when using
$E_k = \tfrac{1}{2}mv^2$
(i.e. $E_k = \tfrac{1}{2}m(v^2)$
not $E_k = (\tfrac{1}{2}mv)^2$!)

Gravitational potential energy

To **lift** an object through a height against gravity we need to **do work**.

The lifting force required is equal in size to the object's weight. $F = W = mg$

The object now has **potential energy**.

$$E_P = E_w$$
$$= F \times d$$
$$= mg \times h$$

$$\boxed{E_P = mgh}$$

(where $g = 10 \, \text{N/kg}$)

A forklift truck has just done work. It has lifted a 50kg crate 5m onto a shelf.

$$E_P = mgh$$
$$= 50 \times 10 \times 5$$
$$= 2500 \, \text{J}.$$

Whenever we lift a mass up through the gravitational field we do work.

The object stores this as **gravitational potential energy**.

Power

Power is the **work done every second** or the **rate of doing work**.
Power is the **energy transferred in unit time**, i.e. how much work is done in each 1s.

$$\text{Power} = \frac{\text{work done}}{\text{time taken}} \qquad P = \frac{E}{t} \qquad \text{Power is measured in watts (W)}$$

Example

A crane is lifting a 10 000 N load to a height of 5 m.

$$W = F \times d$$
$$= 10\,000 \times 5 = 50\,000 \, \text{J or 50 kJ}$$

If the crane can do this in 100 s, its power is:

$$P = \frac{W}{t} = \frac{50\,000}{100}$$
$$= 500 \, \text{J/s or 500 W}$$

Example

A man runs 10 m up a flight of stairs in just 5 s. He weighs 1200 N.

Calculate the work he does and his power.

$$W = F \times d = 1200 \times 10 = 12\,000 \, \text{J or 12 kJ}$$

Remember that the man is doing work due to his vertical motion so it is the vertical distance he moves which is important.

$$P = \frac{W}{t} = \frac{12\,000}{5} = 2400 \, \text{W or 2.4 kW}$$

Quick Test

1. Calculate the work done when a lawnmower is pushed 30 m by a force of 250 N.

2. Calculate the height to which a 500 N load is lifted by a crane if 20 kJ of work is done.

3. Calculate the force that is applied to a car engine if 1500 J of work is done when it is lifted on to a bench 1m high.

4. What is power?

5. What units do we use to measure:
   a) work done?   b) power?

6. Calculate the power of the person pushing the lawnmower in question 1 if the work is done in 25s.

7. Calculate the power of the crane in question 2 if the lift takes 40 s.

8. Calculate the energy to raise 20 kg of water into a tank 7 m high.

9. Calculate the kinetic energy of a runner of mass 100 kg moving at 6 m/s.

Answers 1. 7500J **2.** 40m **3.** 1500N **4.** The rate of doing work **5. a)** Joules **b)** watts **6.** 300W **7.** 500W **8.** 1400J **9.** 1800J

Test your progress

Use the questions to test your progress.
Check your answers at the back of the book on page 108–109.

1. Calculate the speed of a runner who travels 200m in 20s.

   ....

2. Calculate the distance travelled by a motorist who travels at a speed of 90 km/h for three hours.

   ....

3. Calculate the time it would take for a cyclist travelling at 20m/s to travel a distance of 500m.

   ....

4. State how a driver's tiredness affects the total stopping distance of a vehicle.

   ....

5. A girl pushes a trolley with a force of 10N. How much work has she done when the trolley has moved 5m in the direction of the force?

   ....

6. Name three things that may happen to an object that is moving with constant velocity when unbalanced forces are applied to it.

   ....

7. Explain how the shape of a fish helps it to swim easily through water.

   ....

8. Explain why several seconds after jumping from an aircraft a sky diver will be travelling at a constant speed.

   ....

9. Calculate the acceleration of a car which starting from rest reaches a speed of 30m/s after 6s.

   ....

10. The diagram right shows a stationary object on a flat surface.
    a) Name the forces X, Y and Z.
    b) If the object is not moving, what can you say about the forces acting on the object?
    c) If the box has a mass of 3kg and g = 10N/kg, what is the value of the force Y?
    d) Suggest two ways in which the friction acting on the object could be reduced.

    ....
    ....
    ....

11. A man with a mass of 80kg runs up 200 steps in 25s. If the height of each step is 25cm calculate:
    a) the total height climbed by the man;   b) the weight of the man;
    c) the work done by the man;   d) the power of the man.   (g = 10N/kg)

    ....
    ....

12. A pound and a penny fall from a man's pocket.
   a) Which reaches the ground first?   **b)** What is the acceleration of each?

....

13. A mountaineer of mass 80 kg climbs a height of 500m. Calculate gain in the mountaineer's potential energy.

....

14. A car accelerates from 0-60 miles/hour in 10s.
   a) Calculate the acceleration.   **b)** Why cannot this car go faster than 100 mph?

....

15. A skate boarder accelerates down a ramp. After 3s she has reached a speed of 12m/s.
   a) Calculate her acceleration.
   b) Calculate the unbalanced force causing this acceleration if her mass is 50kg.

....

16. A gardener pushes a wheelbarrow at a steady speed for 20m using a force of 400 N.
   a) Calculate the work done by the gardener.
   b) This takes the gardener 40s. What power does the gardener produce?

....

17. An ice sprinter accelerates from rest to 8 m/s in 2s then maintains a steady speed till the end 20s later.
   a) Calculate the acceleration at the start.
   b) What total distance was covered?
   c) How do the pushing force and the friction forces compare at the start and during the steady part of the race?
   d) How can the sprinter reduce friction?

....
....

18. A pupil on an electric scooter travels up a 6 m ramp to a height of 1.2 m.
   a) If the combined mass = 50 kg, calculate the potential energy gained travelling to the top of the ramp.
   b) The scooter takes 8s to travel to the top. Assuming no frictional losses, calculate the average power from the electric scooter.

....

19. A fast car decelerates from 35m/s at rate of 4 m/s^2 for 3s on seeing an obstacle.
   a) What speed does the car reduce to in this time?
   b) What force has to be applied to the 80 kg driver to decelerate the driver with the car?

....

20. A 1000 kg vehicle is travelling at 10 m/s. How much more kinetic energy will it have if it is going at 20 m/s?

....

21. A swimmer dives off a 6 m board. What speed does she enter the water at?

....

Sources of energy

We need energy to get things done.
In our homes, in industry and for transport – energy matters!

Fossil fuels to nuclear energy

Fossil fuels are, at present, our **main sources of energy**.
Coal, oil and gas are called fossil fuels.
Fossil fuels are concentrated sources of energy.
Because fossil fuels take millions of years to form they are
called **non-renewable fuels**.

Fossil fuels cannot be replaced once they are used up, they
are said to be **finite**.

The problems with fossil fuels:
- burning fossil fuels releases **carbon dioxide into the
  atmosphere** which causes a rise in temperature known
  as the **greenhouse effect**;
- burning coal and oil produces **gases** that cause **acid rain**;
- we are using up fossil fuels very quickly.

Nuclear energy has become a main source of energy in
several countries. Nuclear fuel is available and a very concentrated source of energy. Nuclear
waste is however **radioactive** and nuclear fuel is also a **finite** resource.

Dead plants
and animals
are covered
with mud
and earth.

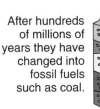

After hundreds
of millions of
years they have
changed into
fossil fuels
such as coal.

Conserving energy

We need to consider how to slow down the rate at which we are using non-renewable fuels.

In the home. Public awareness to
insulate and turn down the heating helps.

Heat escapes:
- by **conduction** through walls
  and floors.
- by **convection** through the roof,
  drawing cold air in through gaps.
- by **radiation** from the walls, windows
  and the roof.

25% through roof, reduced
by putting insulation into loft.

25% through gaps
and cracks around
doors and windows,
reduced by fitting
draft excluders.

10% through
windows, reduced
by installing
double glazing.

25% through
walls, reduced
by having
cavity wall insulation.

15% through floor, reduced by
fitting carpets and underlay.

Heat flows from a hot object to colder surroundings. The heat loss is greater in winter than in
summer as the temperature difference with the surroundings is greater.

In industry. Large buildings need to be insulated and economically heated. As the heat rises to
the roof it should be circulated by large fans to avoid overheating the building.

In transport. Reduce petrol consumption by **car sharing** or using **public transport**. Drivers with
empty seats could be saving
cost as well as energy. The
following shows the order of
efficiency (when full):

most efficient ⟶ least efficient

Renewable and non-renewable sources

Renewable	Advantages	Disadvantages
Water	Free fuel, clean and endless supply.	Requires mountains or large wave area
Geothermal		Few suitable sites, difficult to convert
Solar		Poor in cold climates, only small output
Wind		Variable supply
Biomass	Low-technology	Slow growing

Non-renewable	Advantages	Disadvantages
Coal	Long supply – 300 years	Dirty, pollution and mining damage
Oil	100 years, main fuel for industry	Spillage, finding and extraction
Gas	Clean, easy to pipe for heating and cooking	40 years supply, storage.
Nuclear (uranium)	Small quantities, long supply	Radioactive waste, storage

Energy units

Energy usage can be calculated using: $\boxed{E = P \times t}$

In physics, the basic unit of energy is the **Joule (J)** but around the home the **kilowatt hour (kWh)** is more useful. The kWh is used as a **unit** of electricity and appears on domestic bills.

In industry, where large amounts of energy are supplied at high power we see more prefixes.

$$
\begin{aligned}
1 \text{ kilojoule} &= 1\,\text{kJ} = 1\,000\,\text{J} &&= 10^3\,\text{J} \\
1 \text{ megajoule} &= 1\,\text{MJ} = 1\,000\,000\,\text{J} &&= 10^6\,\text{J} \\
1 \text{ gigajoule} &= 1\,\text{GJ} = 1\,000\,000\,000\,\text{J} &&= 10^9\,\text{J} \\
1 \text{ terajoule} &= 1\,\text{TJ} = 1\,000\,000\,000\,000\,\text{J} &&= 10^{12}\,\text{J}
\end{aligned}
$$

The prefixes will also apply to many other physics units, e.g. $1.2\,\text{GW} = 1\,200\,000\,000\,\text{W}$ of power.

Industry will also use more specialised units such as **tonnes of coal equivalent**, and the **therm**.

Top Tip

Problems in energy supply and demand involve large numbers. Watch out for the prefixes and do not leave these out of your calculations.

Quick Test

1. What are the fossil fuels?
2. What is meant by non-renewable?
3. Why does a house lose heat quicker in winter than in summer?
4. How can industry conserve energy?
5. Name 3 renewable sources of energy.
6. Why might you not be cooking with gas when you retire?
7. If a mega is a million, what is a tera?

Answers 1. Coal, oil and gas **2.** They cannot be replaced once used **3.** Greater temperature difference with the surroundings **4.** Improve insulation and fan hot air down **5.** Water, solar, geothermal, wind **6.** The gas will have run out **7.** A million million

Generation of energy: non-renewables

Thermal power stations

Top Tip
Turbine → Generator → Transformer → National Grid are common to most types of power station. We will study these later, just remember the order for now.

Most of the electricity we use at home is generated at **power stations**. There are several different types of power station but the most common in the UK use **coal** or **gas** as their source of energy (**fuel**).

Fuel is burned to release **energy**.

Heat energy released heats water and turns it to **steam**.

The steam **turns turbines**.

The turbines **turn large generators**, to produce **electrical energy**.

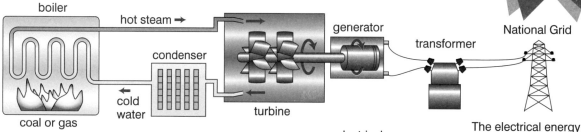

The electrical energy is carried to our homes through the **National Grid**.

chemical energy ▶ heat energy ▶ kinetic energy ▶ electrical energy

Nuclear power stations

In a nuclear power station, a **nuclear reactor** is used instead of the boiler to produce heat. The heat is still used to make steam to drive a turbine and generator just like in thermal power stations.

In the reactor, **nuclear fuel** (uranium) is **bombarded** with **neutrons**. The uranium atom splits into two large parts and two or more neutrons are also released from its nucleus. A large amount of energy is released. This energy drives the power station.

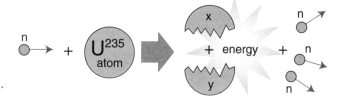

The process of splitting the atom is called **nuclear fission**.

Far less fuel is used in a nuclear power station.
1 kg of coal releases 28 MJ of energy, 1 kg of uranium releases 5 000 000 MJ.

A chain reaction

The neutrons released during fission can set off a **chain reaction**. **Control rods** absorb some of these neutrons.

The control rods are **lowered further** into the reactor to **reduce the energy output** or shut it down.

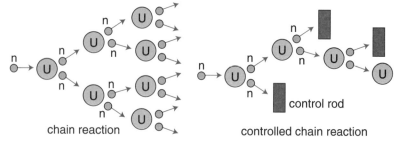

chain reaction

controlled chain reaction

The control rods are **lifted higher out** when the **demand for energy increases**.

Unfortunately, radioactive waste is produced in nuclear reactors.
Some **radioactive waste is very dangerous**.

Efficiency of energy transformation

Energy is always lost to the surroundings in energy converters including power stations. The waste energy is mainly heat. Overall we say energy is being **degraded** because the waste heat can no longer be used to generate electricity.

We want to make power stations (and all devices) as efficient as possible.

Efficiency measures how much of the total we put in is kept as **useful output**.

$$\text{Efficiency} = \frac{\text{Useful Output}}{\text{Total Input}}$$

This gives efficiency as a fraction of 1.

It is more common to give **efficiency** as a **percentage**. This will always be less than 100%.

$$\text{Percentage Efficiency} = \frac{\text{Useful Energy Output}}{\text{Total Energy Input}} \times 100$$

$$\text{Percentage Efficiency} = \frac{\text{Useful Power Output}}{\text{Total Power Input}} \times 100$$

$$P = \frac{E}{t}$$

Combined heat and power

A **thermal power station** is usually no more than **40% efficient**.

A **combined heat and power (CHP)** station uses its **waste heat** to supply hot water for heating local homes and factories. The overall efficiency is much improved.

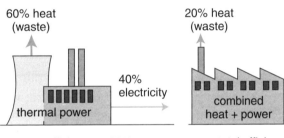

60% heat (waste) — thermal power — 40% electricity — total efficiency = 40%

20% heat (waste) — combined heat + power — 50% heat (used) — 30% electricity — total efficiency = 80%

Pumped hydroelectric

Thermal power stations cannot easily reduce their output in line with the fall in demand at night. Surplus electricity is used, instead of being wasted, to pump water up a height to the reservoir of a hydroelectric scheme. The water now has **gravitational potential energy**. This can be released during the day, when the demand is high, to meet the **peak demand**. The **potential energy** of the water changes to **kinetic energy to drive the turbines**.

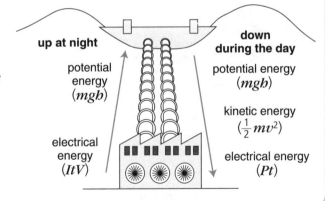

up at night
potential energy (mgh)
electrical energy (ItV)

down during the day
potential energy (mgh)
kinetic energy $(\frac{1}{2}mv^2)$
electrical energy (Pt)

Quick Test

1. What is the energy change in a boiler?

2. How many kilograms of coal would be needed to give the same energy as 1 kg of uranium?

3. What is the main problem with nuclear power stations?

4. Explain a chain reaction.

5. For every 1000 J of energy put in, 650 J are wasted. What is the efficiency of this machine?

Answers 1. chemical to heat **2.** 5 000 000 / 28 = 178 572 kg **3.** radioactive waste **4.** for every neutron absorbed, two or more are released **5.** 35%

Generation of energy: renewables

✔ = advantages, ✗ = disadvantages

Geothermal

In regions where the Earth's crust is thin, **hot rocks beneath the ground** can be used to heat water, turning it into steam. This steam is then used to drive turbines and generate electricity.

✔ **Renewable** source of energy.

✔ No pollution and no environmental problems.

✗ Very few **suitable sites**.

✗ **High cost** of drilling deep into the ground.

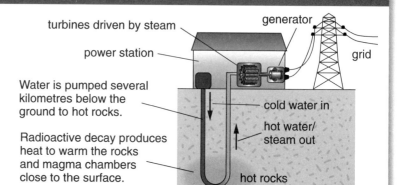

turbines driven by steam

power station

generator

grid

Water is pumped several kilometres below the ground to hot rocks.

Radioactive decay produces heat to warm the rocks and magma chambers close to the surface.

cold water in

hot water/ steam out

hot rocks

Tidal power

At high tide, water is trapped behind a barrage or dam. When it is released at low tide the **gravitational potential energy** of the water changes into **kinetic energy** which then drives turbines and generates electricity.

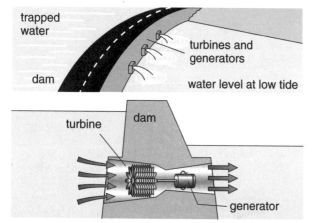

trapped water

turbines and generators

water level at low tide

dam

turbine

dam

generator

✔ **Renewable** source.

✔ **Reliable**: two tides per day.

✔ No atmospheric **pollution**.

✔ **Low** running costs.

✗ **High** initial cost.

✗ Possible **damage to environment**, e.g. flooding.

✗ Obstacle to **water transport**

Top Tip
Don't waste time memorising the diagrams, but study them and remember the advantages and disadvantages of each resource.

Solar energy

The energy carried in the **Sun's rays** can be converted directly into electricity using solar cells.
or
The energy carried in the Sun's rays is absorbed by dark coloured panels and used to **heat** water.

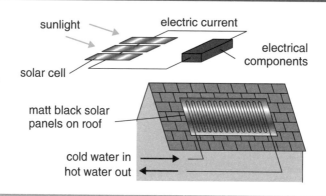

sunlight

electric current

solar cell

electrical components

matt black solar panels on roof

cold water in
hot water out

✔ No pollution.

✗ Initially quite **expensive**.

✗ May not be so useful in regions where there is **limited sunshine**.

Biomass

The **chemical energy** stored in 'things that have grown', e.g. wood, can be **released by burning it**. This energy source can be maintained by growing a succession of trees and then cropping them when they mature.

✔ **Renewable** source of energy.

✔ **Low-level technology**, therefore useful in developing countries.

✔ Does not add to the greenhouse effect as the carbon dioxide released by trees and plants when burned was taken from the atmosphere as they grew.

✗ **Large areas** of land needed to grow sufficient numbers of trees.

Wind power

The kinetic energy of the **wind** is used to drive turbines and generators.

✔ It is a **renewable** source of energy and therefore will not be exhausted.

✔ Has **low-level technology** and therefore can be used in developing countries.

✔ No atmospheric pollution.

✗ **Visual** and **noise pollution**.

✗ Limited to **windy** sites.

✗ No wind, no energy.

Hydroelectricity

The kinetic energy of **flowing water** is used to drive turbines and generators.

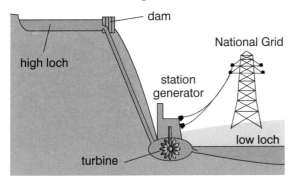

✔ **Renewable** source.

✔ Energy can be **stored** until required.

✔ No **atmospheric pollution**.

✗ **High** initial cost.

✗ **High** cost to environment, e.g. flooding, loss of habitat.

Wave power

The rocking motion of the waves is used to generate electricity.

simple wave machine — the energy in the water waves make this machine rock

this motion is then used to generate electricity

✔ **Renewable** source.

✔ No atmospheric pollution.

✔ Useful for isolated islands.

✗ **High** initial cost.

✗ Visual pollution.

✗ **Poor energy capture**: large area of machines needed even for small energy return.

Quick Test

1. Name three ways in which water could be used as an energy resource.
2. Name two energy resources which may pollute the environment visually.
3. Name two energy resources which could be easily used and maintained in developing countries.
4. Name two energy resources whose capture requires a suitable site that might be rare.
5. Name one energy resource whose capture might cause audible pollution.

Answers 1. Hydroelectricity, tidal, wave **2.** Wind, wave **3.** Wind, biomass **4.** Geothermal, tidal **5.** Wind

Electromagnetic induction

Inducing voltage and current

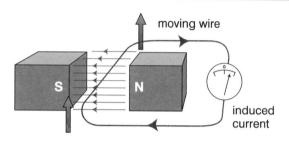

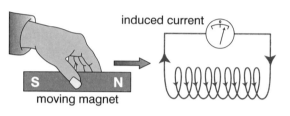

Wire

- If a **wire** is **moved across a magnetic field**, a **voltage is induced** across the wire.
- If the wire is part of a **circuit**, a **current will flow**.
- If the wire is moved in the **opposite direction**, the **induced voltage** and **current** are in the **opposite direction**.
- If the wire is held **stationary**, **no current** or **voltage** is induced.
- To increase the size of induced voltage or current, we could:
  - **a) increase** the **field strength** (use a stronger magnet);
  - **b)** move the wire **more quickly**.

Coil

- If a **magnet is moved into a coil**, a **voltage and current is induced** in the coil.
- If the **magnet is pulled out**, the induced voltage and current is in **the opposite direction**.
- If the coil is in a **changing magnetic field**, a **voltage is induced in the coil**.
- To increase the size of the induced voltage and current, we can:
  - **a) increase** the **field strength** (use a stronger magnet);
  - **b) increase** the magnet's **speed**;
  - **c)** put **more turns** on the coil.

Current and voltages can be created when magnetic field lines are cut by a conductor, e.g. a piece of wire, or when a coil is in a **changing magnetic field** e.g. near a moving or spinning magnet. Producing currents and voltages in this way is called **electromagnetic induction**.

The simple dynamo

A **dynamo** like that used on a bicycle is used to generate small currents.

- As the wheel **rotates** it turns the knurled knob.
- The magnet and its magnetic field **spin around**.
- Its **magnetic field lines cut through the coil inducing a current in it**.
- The current generated **keeps changing** size and direction.
- This is called an **alternating current**.
- This current can be used to light the bicycle's lights.
- If the cyclist stops, the **wheel stops**. There is **no movement** of the magnet and its field, so there is **no induced current** and the light will go out.

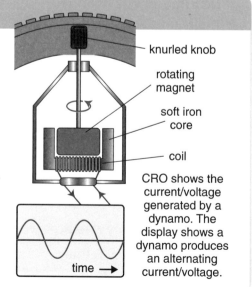

knurled knob

rotating magnet

soft iron core

coil

CRO shows the current/voltage generated by a dynamo. The display shows a dynamo produces an alternating current/voltage.

time →

Generators and alternators

Model AC generators

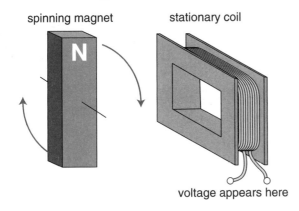

spinning magnet   stationary coil

voltage appears here

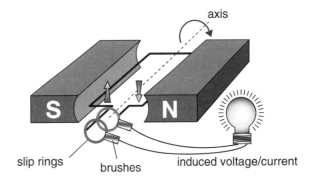

axis

S   N

slip rings   brushes   induced voltage/current

The **coil** is in a **changing magnetic field**, so a **voltage is induced in the coil**.

This model could be used in a small boat to measure its speed, with the magnet part of a spinning propeller under the hull.

If a **coil is rotated** between the poles of a magnet a **current is induced** in the coil.

Because the wires are continually **changing direction** as they rotate, the induced current also changes size and direction.

The induced current is an **alternating current**.

A **generator** which produces alternating current is called an **alternator**.

The coil will generate a larger current if a) a **stronger magnet** is used, b) the coil is **turned more quickly** or c) a coil with **more turns** is used.

A full sized generator

A larger, commercial generator will have the spinning magnet replaced with **electromagnets**. **Several coils** can be used.

These will be supplied with current by a small dynamo.

These spinning electromagnets are called the **rotor coils**.

The power is induced in the external **stator coils** (several and stationary).

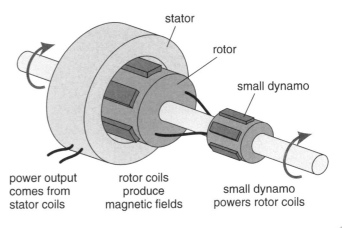

stator

rotor

small dynamo

power output comes from stator coils   rotor coils produce magnetic fields   small dynamo powers rotor coils

---

Quick Test

1. What will happen to the current induced in a wire if:
   a) the wire is moved more quickly through the field?
   b) the wire is held stationary in the field?
   c) the wire is moved parallel to the field?
   d) the wire is moved at the same speed but in the opposite direction?

2. What kind of current is produced when a magnet is rotated inside a coil?

3. Suggest two ways in an alternator can be changed to produce a higher current?

4. State two differences a commercial generator will have over a simple model generator.

Answers 1. a) Larger current **b)** no current **c)** no current **d)** same current as in a) but in opposite direction **2.** Alternating current **3.** Increase the speed of rotation, increase the strength of the magnetic field, increase the number of turns on the alternator's coil **4.** Electromagnets instead of spinning magnets, several coils used, power induced in several stator coils.

Transformers and the grid

Transformers are used to change voltages; but only alternating voltages. A transformer consists of two coils wrapped around a soft iron core.

How a transformer works

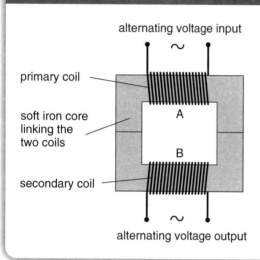

alternating voltage input

primary coil

soft iron core linking the two coils

secondary coil

alternating voltage output

- An **alternating voltage** is applied across coil **A**, called the **primary coil**.
- It produces a magnetic field that is **continuously changing**.
- This changing magnetic field cuts through the coils of the **secondary coil (B)**.
- An **alternating voltage is induced** across the secondary coil.

Top Tip

It is the changing magnetic field that is important to the secondary coil. The iron core is only there to increase the effect by concentrating the field.

How large is the induced voltage?

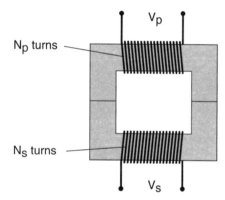

V_P

N_P turns

N_S turns

V_S

The size of the voltage (V_S) induced in the secondary coil depends upon the size of the voltage across the primary coil (V_P), the number of turns on the primary coil (N_P) and the number of turns on the secondary coil (N_S). They are linked by the equation:

$$\frac{V_S}{V_P} = \frac{N_S}{N_P}$$

A transformer like this which is used to increase voltages is called a **step-up transformer**. If the transformer decreases voltages it is called a **step-down transformer**.

Example

Calculate the voltage across a secondary coil of a transformer if an alternating voltage of 12 V is applied across the coils of the primary. The number of turns on the primary coil is 100 and the number of turns on the secondary coil is 500.

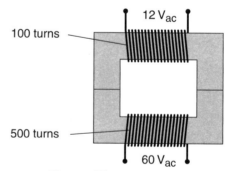

12 V_{ac}

100 turns

500 turns

60 V_{ac}

Using $\quad \dfrac{V_S}{V_P} = \dfrac{N_S}{N_P}$

$$V_S = \frac{N_S}{N_P} \times V_P$$

$$V_S = \frac{500}{100} \times 12 = 60\,V$$

Efficiency of a transformer

In an **ideal transformer**, no power is lost:

Input Power = Output Power $V_P I_P = V_S I_S$ so $\dfrac{V_S}{V_P} = \dfrac{N_S}{N_P} = \dfrac{I_P}{I_S}$

In a **step-up transformer**, the **voltage is stepped up**, and the **current is stepped down**.

In a **step-down transformer**, the **voltage is stepped down**, and the **current is stepped up**.

In a **real transformer**, power is lost as heat and sound:
- the coils heat up;
- the iron core heats up;
- sound energy emits from the core.

The output power will be less than 100% of the input power.

The National Grid

Electrical energy is carried by long **transmission lines** between the power station and our homes and industries.

Long transmission lines have **resistance** and there is power loss in the lines.

$$P_{loss} = I^2 R$$

This loss is kept to a minimum by **increasing the voltage** using a step-up transformer so that the **current is decreased**. This is why the **voltage in power lines** is always kept **high**. Near our homes the **voltage is decreased to a safe level** by a step-down transformer.

Quick Test

1. What kind of voltages and currents can a transformer change?

2. What kind of transformer:
   a) increases voltages?
   b) decreases voltages?

3. Name 3 ways in which energy is lost from a transformer.

4. Calculate the voltage across the secondary coil of the transformer if an alternating voltage of 20 V is applied across the coils of the primary. The primary has 400 turns on the coil and the secondary has 100 turns on the coil.

5. Why is electrical energy transmitted through the National Grid at high voltages and low currents?

6. Why is the voltage of the electrical supply decreased for supplies to houses?

Answers 1. Alternating (a.c.) **2. a)** step-up **b)** step-down **3.** heat in the coils, heat in the core, sound **4.** 5V **5.** To reduce energy loss **6.** Low voltages are safer.

Heat capacity

Heat has the capacity to change the **temperature** or the **state of a substance.**

Heat and power

Power is the rate of transferring energy.

$$P = \frac{E}{t}$$

The amount of (heat) energy supplied,

$$E = P \times t$$

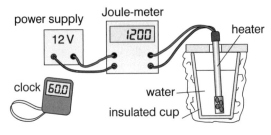

power supply 12 V — Joule-meter **1200** — heater — clock **60.0** — water — insulated cup

e.g. A low voltage immersion heater is supplying energy to heat water in a cup.

$$P = \frac{E}{t} = \frac{1200}{60} = 20\,W.$$

This heater has a power of 20 W.

Heat and temperature

Heat is the **energy supplied.**   **Temperature** is how hot an object becomes.

The amount of heat energy required to raise the temperature of a substance:
- varies with the **change in temperature** required ($E \propto \Delta T$);
- varies with the **mass being heated** ($E \propto m$);
- varies with the **specific heat capacity** of the material being heated ($E \propto c$).

Note: The **greater the temperature change** we desire, the **more energy will be required** or given out. A **different mass of material** will require a **different amount** of heat for the same temperature rise. The same mass of **different materials** requires **different quantities of energy** to raise their temperature by one degree.

The amount of heat energy depends on all three factors:

J/kg°C ⟶ kg
J ⟶ $E = cm\Delta T$ ⟵ °C

$$c = \frac{E}{m\Delta T} \qquad m = \frac{E}{c\Delta T} \qquad \Delta T = \frac{E}{cm}$$

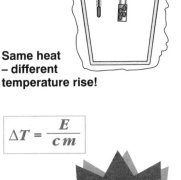

thermometer — identical heaters

Same heat – different temperature rise!

Specific heat capacity

Specific heat capacity, c, is the amount of energy required to change the temperature of **1 kg of a substance by one degree**. This is also the energy 1kg of a substance can store for each degree.

Water has a very high specific heat capacity, c = 4180 J/kg°C. It takes a lot of heat to make water hot. Water can store a lot of heat.

Top Tip Remember heat and temperature are different ideas.

Cooling curves

When the temperature drops, energy is being given out.

When the temperature stays the same, energy is still being given out! A change of state occurs.

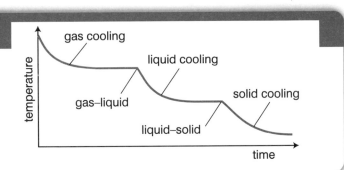

temperature vs time: gas cooling, gas–liquid, liquid cooling, liquid–solid, solid cooling

Heat and change of state

Solid state: Atoms **vibrate**.

Liquid state: Atoms have **more energy** and are **free to tumble**.

Gas state: Atoms have **enough energy to break free and stay apart**.

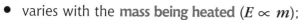

solid   liquid

gas

Energy is gained or lost by a substance when its state is changed. A change in state does **not** involve a change in temperature.

The amount of heat energy required to change the state of a substance:

- varies with the **mass being heated** ($E \propto m$);
- varies with the **specific latent heat** l of the material being heated ($E \propto l$).

Note: A different mass of material will require a **different amount of heat** to change its state. The same mass of **different materials** requires **different quantities of energy** to change their states.

The amount of heat energy depends on these two factors:

$$J \xrightarrow{\hspace{0.5cm} kg} \boxed{E = ml} \searrow_{J/kg} \qquad \boxed{l = \frac{E}{m}} \qquad \boxed{m = \frac{E}{l}}$$

Specific latent heat

The specific latent heat, l, is the amount of heat energy to change the state of 1kg of a substance.

Latent heat of fusion – the amount of energy taken in or released to change between a **solid** and a **liquid state**.

e.g. specific latent heat of fusion of ice, $l = 3.34 \times 10^5$ J/kg.

Latent heat of vapourisation – the amount of energy taken in or released to change between a **liquid state** and a **gas**.

e.g. specific latent heat of vaporisation of water, $l = 2.26 \times 10^6$ J/kg.

Energy changes

The principle of conservation of energy allows us to calculate the heat energy supplied to a substance from an electrical heater.

Heat energy gained = Electrical energy used. $\boxed{cm\Delta T = ItV}$ or $\boxed{ml = ItV}$

Quick Test

1. If you heat 1 kg of water in a kettle from room temperature of 20°C to boiling point of 100°C, how much energy does this take?

2. How long would this kettle take to boil if its power is 2200W?

3. In practice, this kettle took 3 minutes. Why?

4. After boiling the water, this kettle is left on. How much more energy is used while all the water turns to steam?

5. How long does the water take to evaporate?

Answers 1. $E = cm\Delta T = 4180 \times 1 \times 80 = 334\,400$ J **2.** $t = E/P = 334\,400 / 2200 = 152$ s **3.** Heat was lost to the surroundings **4.** $E = ml = 1 \times 2.26 \times 10^6 = 2.26 \times 10^6$ J **5.** $t = E/P = 1027$ s, ~ 18 minutes.

Test your progress

Use the questions to test your progress.
Check your answers at the back of the book on pages 108–109.

1. Name a device which changes:
   a) electrical energy into kinetic energy   b) kinetic energy into electrical energy
   c) light energy into electrical energy

   ....

2. What kind of energy does a lift have when it is stopped at the top of a building?

   ....

3. Name the main method of heat transfer in each of the following:
   a) Sun to Earth   b) radiator to room   c) cooker hot-plate to food in saucepan.

   ....

4. a) Explain what is meant by the phrase 'renewable energy'.
   b) Name three renewable sources of energy.
   c) Name three non-renewable sources of energy.

   ....

   ....

   ....

5. Suggest three ways of reducing the rate at which fossil fuels are used.

   ....

6. What steps could be taken to reduce heat-loss from a house.

   ....

7. A crane used 500J of electrical energy to give a crane 300J of gravitational potential energy. Calculate the efficiency of the crane.

   ....

8. An electric light bulb is 20% efficient. If 200J of electrical energy enters the bulb, how much light energy is produced?

   ....

9. The diagram (right) shows a hydroelectric power station.
   a) What kind of energy does the water possess in the top loch?
   b) What kind of energy does the water possess as it enters the turbine?
   c) Explain how the surplus energy could be stored until it is needed.
   d) Calculate the efficiency of a turbine which transfers 800kJ of energy into 600kJ of electrical energy every second.

   ....

   ....

10. In a house, 45% of heat is lost through the walls and floors, 20% to draughts and 10% through the windows. Where is the rest lost and what percentage is lost there?

....

11. If 50J of energy can pass through 1m² of double glazing every second, how much heat escapes every second from a window of area 5m² ?

....

12. A bulb in a playhouse is to run from a 11.5V supply. A transformer is used to step down the mains from 230V. If the primary coil has 1000 turns, how many turns has the secondary coil?

....

13. Why can a transformer not be used with a battery?

....

14. When a material changes from a solid to a liquid, what happens to the temperature of the material?

....

15. If a whole house contains 400 kg of air, how much energy will it take to raise the temperature of the house by 15°C? (c_{air} = 1000J/kg °C.)

....

16. What is the purpose of the National Grid?

....

17. State the difference between heat and temperature.

....

CREDIT

18. A beaker containing 0.25 kg of water is to be heated till its temperature changes from 15 to 30°C. A 50W immersion heater is placed in the water and the beaker is wrapped in cotton wool and metal foil.
a) What minimum time is required to heat this water to 30°C? (c_{water} = 4180J/kg°C.)
b) What is the purpose of the two wrappings?

....

19. 0.1 kg of ice is placed in a jug of juice. How much energy is removed from the juice as the ice melts? (Specific latent heat of fusion of water = 3.34×10^5 J/kg)

....

20. Industrial power tools are often designed to operate at 110V. They require a transformer to be used with the mains.
a) If one such tool draws a current of 16A from the transformer, what current might be expected from the mains?
b) In fact the mains supplies a current of 12A to the transformer. What is the efficiency of the transformer?

....

21. Why, in practice, are transformers not 100% efficient?

....

Viewing the universe

The Solar System

Our Solar System consists of a **star, a number of planets, moons, asteroids and comets.** We call **our star the Sun.** It contains over 99% of the mass in our Solar System. The planets, their moons, the asteroids and comets all orbit the Sun.

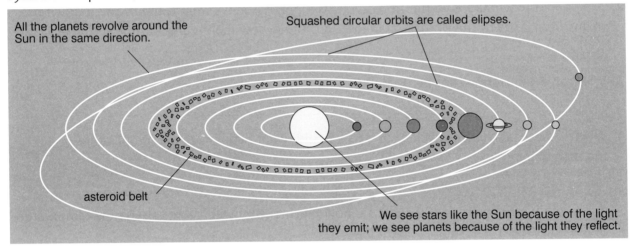

All the planets revolve around the Sun in the same direction.

Squashed circular orbits are called elipses.

asteroid belt

We see stars like the Sun because of the light they emit; we see planets because of the light they reflect.

The earth is one of eight **planets.** You do not need to learn their names but you may wish to! Mercury, Venus, Earth, Mars, Jupiter, Saturn, Uranus and Neptune.

You can write a memory aid: **M**y **V**ery **E**xcellent **M**um **J**ust **S**erved **U**s **N**achos.
4 small and rocky, the asteroid belt, 4 giant planets of gas, Pluto is a dwarf planet.

Key Fact

Planet	a natural satellite of the Sun.
Moon	a natural satellite of a planet.
Sun	a star (1 of 100 000 million in our galaxy!)
Star	emits light and heat radiation.
Solar system	the sun and the objects that orbit it.
Galaxies	consists of millions of stars.
Universe	contains billions of galaxies.

Top Tip
Learn the key terms of the Solar System.

Stars and galaxies

- Our Sun is a **star.**
- It is one of millions of stars that makeup the **galaxy** we live in.
- Our galaxy is called the **Milky Way.**
- In the **Universe** there are billions of galaxies. They are separated by distances that are often millions of times greater than the distances between stars in a galaxy.

top view of our galaxy

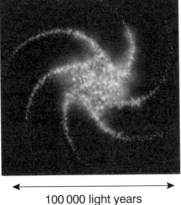

100 000 light years

side view of our galaxy

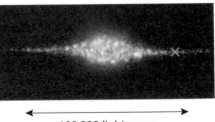

100 000 light years

The Milky Way is a spiral galaxy
✕ we are about here

Distances – light years

Distances are so vast in space that astronomers use a new unit for distance.
1 light year: The distance travelled by light in one year. This is not a time!
You can calculate how far this is in metres using $d = v \times t$.

(Hint: Work out how many seconds in a year to work out a light year.)
$d = v \times t = (3 \times 10^8) \times (1 \times 365 \times 24 \times 60 \times 60) = 9.46 \times 10^{15}\,\text{m}$

Here are some approximate values for distances. You need to learn these:

Earth to Moon	1.2 light seconds
Earth to Sun	500 light seconds (8 minutes)
Earth to next star	4.3 light years
Across our galaxy	100 000 light years.

The refracting (astronomical) telescope

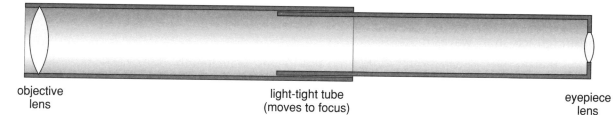

objective   light-tight tube   eyepiece
lens   (moves to focus)   lens

Objective lens: Has a large diameter to collect many light rays for a brighter image. Produces an inverted image inside the tubes.

Eyepiece lens: Magnifies this image. Here is how:

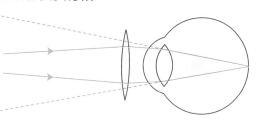

The image being magnified is close to the lens. The light rays appear to come from far apart. The combined result is virtual, inverted and magnified.

> **Top Tip**
> Know the 3 main parts of the telescopes and what they do.

Astronomers also use **reflecting telescopes** (using mirrors to bend light paths) which can be made bigger and so give a brighter image.

Quick Test

1. Name the parts of the universe in order, biggest first.

2. Change to metres:
   a) 1 light minute (distance travelled by light in 1 minute)
   b) 1 light second (the distance travelled by light in 1 second).

3. Why are some telescopes called refractors?

4. What is the purpose of the objective and the eyepiece lenses?

Space physics

Exploring the spectrum

Dispersion of light

White light can be split (**dispersed**) into its different colours using a prism.

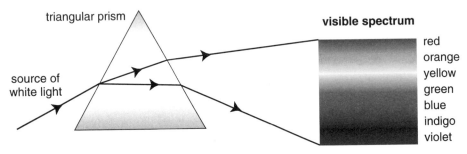

Easy to remember: **R**ichard **O**f **Y**ork **G**ave **B**attle **I**n **V**ain.

Different colours of light correspond to different wavelengths.

The prism refracts short waves most.

Violet has the shortest wavelength.   Red has the longest wavelength.

Spectroscopy

A **spectroscope** can also be used to view spectra. An image is made in the eye.
A **spectroscope** has a slit at one end and an eyepiece lens at the other.

Continuous spectra

The Sun (and light bulbs) emit a continuous spectrum containing all the wavelengths of visible light.

Line spectra

Some light sources (e.g. glowing gases) emit only certain wavelengths. They produce a line spectrum.

Top Tip
A line spectrum is a fingerprint of atoms and the elements of a star.

Light comes from within the structure of the atom. **Each atom has its own structure and line spectrum pattern**. Astronomers look at light spectra to identify elements present in distant stars.

96

The electromagnetic spectrum

short wavelength
high frequency

long wavelength
low frequency

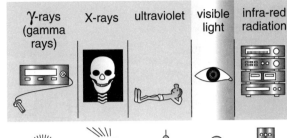

| γ-rays (gamma rays) | X-rays | ultraviolet | visible light | infra-red radiation | microwaves television | radio |

| radioactive material | x-ray machine | u.v. lamp | light bulb | remote control | transmitter mast |

The electromagnetic spectrum is a family of waves with a large number of common properties:

- They are all able to travel through a vacuum.
- They all travel at the same speed through a vacuum i.e. the speed of light ($300\,000\,000$ m/s).
- They all transfer energy.
- They can all be reflected, refracted, and diffracted.

Some of the properties of these waves change as the wavelength and frequency change. The family is therefore divided into smaller groups shown above.

Looking into space

Stars and galaxies emit radiation at many different frequencies in the electromagnetic spectrum.

The atmosphere absorbs many of these.

The Hubble Telescope in orbit around the Earth has been able to see 10 times further out into space.

The Chandra Observatory finds information using X-rays from objects in space.

The Spitzer Telescope is seeing more detail in the region of infra-red.

The radio telescope

Radio waves from space travel through the atmosphere.
Radio telescopes can tune into radio waves from exploding stars and exploding galaxies.
Radio waves have a very long wavelength.
Radio telescopes have very large dishes.
The aerial at the focus changes radio waves into an electric signal.
Radio telescopes can be joined together to make the signal stronger.

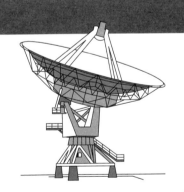

Quick Test

1. Which colour refracts the least? What is its wavelength?

2. What do line spectra tell us about?

3. What type of radiation is visible light?

4. List the EM spectrum in order of increasing wavelength.

Answers 1. Red, 700 nm **2.** What atoms are present **3.** Electromagnetic **4.** Gamma, X-rays, ultra-violet, visible, infra-red, microwave, TV, radio.

Rockets and gravity

Acceleration and flight

Forces occur in equal and opposite pairs. You push on a wall and the wall pushes you away. But how can a rocket in space push off? The rocket exerts a thrust on its burning fuel, the fuel exerts an equal but opposite thrust on the rocket. (Thrust is another name for a force).

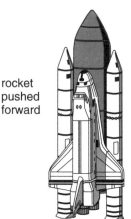

rocket pushed forward

Newton's Laws

Newton's laws are used throughout space travel:

Newton's 3rd Law

> If A exerts a force on B, then B exerts an equal but opposite force on A.

NIII tells us forces exist in **pairs**.

Top Tip
You should now know, and be able to use, Newton's three laws of motion.

fuel pushed back

Newton's 2nd Law

> $F = ma$ or $a = \dfrac{F}{m}$

Thus the fuel pushing on the rocket causes acceleration.
NII tells us a force causes **acceleration**.

Newton's 1st Law

> An object will remain at rest, or stay at constant speed in a straight line, unless acted on by an unbalanced force.

Thus once a rocket has escaped from a planet it can shut off its engines. Interplanetary flight takes place at a constant speed using no fuel.

NI tells us we need **no force** for an object to keep going.

Momentum

A capsule pushes off from its more massive parent ship.

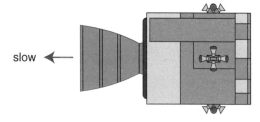
slow

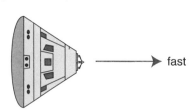
fast

Both experience the same size of force.
The product of **mass** × **velocity** is the same for each.
The capsule has a **smaller mass** × a **larger velocity**.
The ship has a **larger mass** × a **smaller velocity**.

The product is called momentum. | **Momentum** = $m \times v$

Gravitational field strength

Around planets we feel the force of gravity. The larger the mass of the planet the stronger the gravity.

Gravity or **gravitational field strength** is defined as the force on a unit mass.

$$g = \frac{F}{m}$$

On Earth, gravitational field strength is 10 N of pull on every 1 kg of mass.

On Earth, g = 10 N/kg

On the Moon, g = 1.6 N/kg.
On Mars, g = 3.7 N/kg.

To find our weight, our force on a planet, we need to multiply our mass by g.

$$W = mg$$

Suppose your mass is 50 kg.

$W_{Earth} = mg = 50 \times 10 = 500$ N.
$W_{Moon} = mg = 50 \times 1.6 = 80$ N.

(Note: gravitational strength decreases with height as we move away from the planet.)

weight loss, but still the same mass!

Acceleration due to gravity

The force of gravity near a planet's surface gives **all objects the same acceleration down**.

Both accelerate down together.

$$a = 10 \, \text{m/s}^2$$

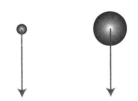

(On the Moon even a feather will follow this rule as there is no atmosphere to give resistance.)

As the force down is equal to our weight $a = \frac{F}{m}$ and $g = \frac{F}{m}$ are equivalent.

On Earth, acceleration, $a = \frac{F}{m} = 10 \, \text{m/s}^2$ and gravitational field strength, $g = \frac{F}{m} = 10$ N/kg.

Top Tip
Acceleration due to gravity depends on gravitational field strength.

Quick Test

1. State Newton's 3 laws of motion.
2. What is momentum the product of?
3. What is your weight on Mars if you're 50 kg on Earth?
4. What would be the acceleration of an object near the surface of the Moon?

Answers 1. Learn from above 2. Mass and velocity 3. W = mg = 50 × 3.7 = 185 N 4. a = g = 1.6 m/s²

Re-entry!

Weightlessness

- At a great distance from a planet the gravity is small and negligible – we can say we are weightless.
- If you were between the Earth and the Moon you would also be weightless where the two gravitational fields cancel each other out.
- The Space Station in orbit is constantly falling to earth. Astronauts inside are falling at the same rate as the Space Station. These free-fall astronauts appear to be weightless.

Projectile motion

Consider a space vehicle before it re-enters our atmosphere.
It is travelling **horizontally** at a **constant speed**.
Gravity pulls on the space vehicle in a **vertically downwards direction**.
This creates an **acceleration downwards**.

The vehicle follows a projectile path.

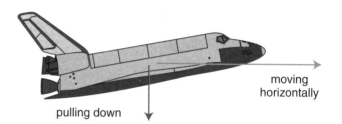

moving horizontally

pulling down

Constant horizontal speed.
$$v = \frac{d}{t}$$

Increasing vertical speed.
$$v = u + at$$

The **resultant speed** is a combination of the **horizontal** and the **vertical**.

Satellite motion

Satellite motion extends the ideas of projectile motion. It was predicted by Newton's thought experiment over 300 years ago.

If a cannon ball is fired off a very high mountain fast enough it will never reach the ground, as Earth is curved. Instead it will remain in orbit and free-fall.

Today satellites are launched into orbit in space at great height.

e.g. low orbit = 200 km,
     geo-stationary orbit = 36 000 km.

The satellites remain in orbit because of gravity.

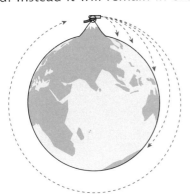

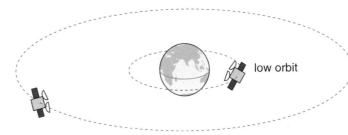

low orbit

geo-stationary orbit

Frictional heating

On re-entry the astronauts start to lose the feeling of weightlessness.

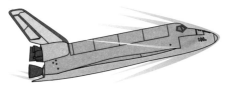

A space craft re-enters the thin atmosphere so fast there is a large frictional force which slows it down.

Because the effect of friction is to change the kinetic energy into heat, the temperature of the outside of the returning space craft is usually well over 1000°C.

Loss in kinetic energy = Heat produced due to work done against friction.

$$E_k \text{ loss} = E_w \text{ done}$$
$$\tfrac{1}{2} m v^2 = F \times d$$

This friction causes the temperature of the protective tiles to rise.

$$E_k \text{ loss} = E_h \text{ gain}$$
$$\tfrac{1}{2} m v^2 = c \, m \, \Delta T$$

Top Tip
Now is a good time to revise the energy equations:
$Ek = \tfrac{1}{2}mv^2$ (kinetic),
$Ew = Fd$ (work) and
$Eh = cm\Delta T$ (heat)

Quick Test

1. What sensation do you experience in free-fall?
2. In which direction does a projectile have:
   a) constant speed?
   b) acceleration?
3. How can a satellite be projected into orbit?
4. What is the energy change of a returning spacecraft?

Answers 1. Weightlessness 2. a) horizontal b) vertical 3. high and fast 4. kinetic to heat

Test your progress

Use the questions to test your progress.
Check your answers at the back of the book on page 108–109.

1. What forces hold planets in orbit?

   ....

2. a) What is a satellite?
   b) Give an example of a natural satellite.
   c) Give an example of an artificial satellite.

   ....

   ....

   ....

3. Which colour is reflected most when white light passes through a prism? What is the name of the band of colours produced by the prism?

   ....

4. How long does light take to travel from the Sun to Earth?

   ....

5. How long does light take to travel across our Galaxy?

   ....

6. Name three types of bodies in our Solar System.

   ....

7. What is the name of all space and stars?

   ....

8. What is a planet?

   ....

9. Which colour of light has the highest frequency?

   ....

10. Which colour of light has the longest wavelength?

    ....

11. What are the names and purposes of each of the two lenses in an astronomical telescope?

    ....

12. Name a type of wave, other than light, which can be detected from space.

    ....

13. What does a line spectrum from a star tell us?

    ....

14. What apparatus can be used to split starlight into its colours?

    ....

15. What are the upwards and downwards forces on a rocket at lift-off called?

....

16. Why does a rocket have an acceleration at lift-off?

....

17. An astronaut and his spacesuit have a combined mass of 160 kg on Earth. What is the weight on the Moon? (g = 1.6N/kg)

....

18. What is meant by the term 'light-year'?

....

19. How can the objective lens of a telescope affect the brightness of the image?

....

20. What is the order of the electromagnetic spectrum?

....

21. What is Newton's Third law?

....

22. What happens to a spacecraft at re-entry to the Earth's atmosphere?

....

Sample Questions

General Level

The General Level exam lasts 1 hour 30 minutes and contains a mixture of question types. Some are Knowledge and Understanding (KU) and some are Problem Solving (PS).

1. **Correct response questions.** Try to eliminate any answers that seem obviously wrong as well as selecting or calculating the correct answer.

    A student is looking at this house.
    How does the image appear on the student's retina?

    A   B   C   D   E

    (You can eliminate D, E and C, as they are only partially inverted. Then recall the image the eye needs to be upside down and laterally inverted, so the answer is A).   [KU 1 mark]

2. **Write a short statement.** This type of question will expect you to be brief but factual with your answer.

    Two-way radios can be used to communicate at distances of up to about 3km.
    Each contains a transmitter and receiver.
    a) At what speed do the signals travel between these radios?
    b) Explain why no cables are required between these radios.

    a) 3×10^8 m/s   [KU 1 mark]
    b) They use radio waves which can travel through air.   [PS 2 marks]

3. **Write a few sentences giving a description or explanation.** A more extended answer than a short statement is required.

    Explain the operation of a stethoscope.

    The bell gathers sound waves from the patient's body.
    The rubber tubing transmits sound from the bell to the earpieces.
    The earpieces transfer the patient's sound to the doctor's ears.   [KU 3 marks]

4. **Do a calculation.** You should try to give formulae and workings as some marks can be awarded for these even if you make a mistake and the answer is wrong. Do not forget to add the units or you will lose half a mark.

    A gondola is carrying skiers up a mountain.
    The mass of the loaded gondola is 5000 kg.
    Calculate the weight of the loaded gondola.

    Give the formula: $W = mg$
    Identify the quantities: $= 5000 \times 10$
    Do the calculation: $= 50\,000\,N$ – add units   [KU 2 marks]

    How much potential energy has the loaded gondola gained?

    Give the formula: $E_p = mgh$
    Identify the quantities: $= 5000 \times 10 \times 2000$
    Do the calculation: $= 100\,000\,000\,J$

    Here you also had to select the correct distance to use.   [PS 2 marks]

Credit Level

The Credit Level exam last 1 hour 45 minutes. It contains a mixture of question types, but does not contain multiple-choice questions. Remember to check whether the question is asking you to **state**, **describe** or **calculate**. As a guide to how much to write, check how many marks the question has been allocated – 2 marks will require more than 1 statement!

1. Radio signals are sent at a frequency of 6 GHz to a satellite in geo-stationary orbit for worldwide transmission.
    a) State what is meant by a geo-stationary orbit.   [KU 2 marks]
    b) Calculate the wavelength of these waves.   [PS 3 marks]
    c) Complete the following diagram to show how a ground station uses a reflector to capture the microwaves.   [KU 2 marks]

State: a) *A geo-stationery satellite stays at the same point above the Earth's surface.*
Calculate: b) $\lambda = v/f = (3 \times 10^8)/(6 \times 10^9) = 0.05\,m$
Complete: c)

2. A hedge cutter has a label displaying electrical information:
    Model JT47-004   230 V ac   50 Hz   1200 W   ▣
    a) State why only two cables are needed in the flex.   [KU 1 mark]
    b) State what colours these two wires would have.   [PS 2 marks]
    c) State what fuse value you would expect to find in the plug.   [KU 1 mark]
    d) A typical person has a resistance through the body from hand to foot of approximately 5000 Ω. If the hedge cutter cut the flex and a current passed through the person to earth, calculate the size of this current.   [KU 2 marks]
    e) Explain why the fuse in the plug would not protect the person in **d)**.   [PS 2 marks]
    f) What is the purpose of the fuse in the plug?   [KU 1 mark]
    g) State and explain why it would be more dangerous for this fault to occur when it is raining.   [PS 2 marks]

State: a) *The appliance is double insulated.*
State: b) *Brown and blue*
State: c) *13 A fuse*
Calculate: d) *I = V/R = 230/5000 = 0.046 A = 46 mA.*
Explain: e) *The current is much less than 13A, so it would not blow.*
State: f) *The fuse protects the flex from overheating.*
State and explain: g) *The current would increase because the water would reduce the resistance.*

Remember – don't panic, there are no trick questions. You have studied for your exam and have practised questions. Ask yourself – what is the question asking me to do? What physics can I use here? Show workings clearly and neatly, and remember to add units.

S grade formulae

Telecommunications

$v = \dfrac{d}{t}$ $f = \dfrac{n}{t}$ $v = f\lambda$ $f = \dfrac{1}{T}$ *CREDIT*

Using electricity

CREDIT

$V = IR$ $I_S = I_1 = I_2 = I_3$ $Q = It$

$R = \dfrac{V}{I}$ $V_S = V_1 + V_2 + V_3$ $I = \dfrac{Q}{t}$ $R_S = R_1 + R_2 + R_3$

$P = VI$ $I_P = I_1 + I_2 + I_3$ $P = I^2 R$ $\dfrac{1}{R_P} = \dfrac{1}{R_1} + \dfrac{1}{R_2} + \dfrac{1}{R_3}$

$E = Pt$ $V_P = V_1 = V_2 = V_3$ $\left(\text{voltage} = \dfrac{\text{energy}}{\text{charge}}\right)$

Health physics

$v = f\lambda$ $A = \dfrac{n}{t}$ $P = \dfrac{1}{f}$ $f = \dfrac{1}{P}$ *CREDIT*

Electronics

$\text{Voltage gain} = \dfrac{V_{\text{out}}}{V_{\text{in}}}$ $\text{Power gain} = \dfrac{P_{\text{out}}}{P_{\text{in}}}$ $P = \dfrac{V^2}{R}$ *CREDIT*

Transport

$\bar{v} = \dfrac{d}{t}$ $F = ma$ $W = mg$ $E_w = Fd$ $P = \dfrac{E}{t}$ $E_P = mgh$

$a = \dfrac{v - u}{t}$ $E_k = \dfrac{1}{2}mv^2$ *CREDIT*

Energy matters

$\dfrac{n_S}{n_p} = \dfrac{V_S}{V_p}$ $E = ItV$ $E_h = cm\,T$

$\% \text{ efficiency} = \dfrac{\text{useful } E_o}{E_i} \times 100\%$ $V_P I_P = V_S I_S$ *CREDIT*

$\% \text{ efficiency} = \dfrac{\text{useful } P_o}{P_i} \times 100\%$ $\dfrac{n_S}{n_p} = \dfrac{I_p}{I_S}$ $\dfrac{V_S}{V_P} = \dfrac{n_S}{n_P} = \dfrac{I_P}{I_S}$ $E_h = m\,l$

Space physics uses formulae from the other topics.

Quantities and units

Quantity	Symbol	Unit	Unit symbol
time	t	second	s
mass	m	kilogram	kg
distance, length	d	metre	m
area	A	square metre	m^2
volume	V	cubic metre	m^3
charge	Q	coulomb	C
current	I	ampere	A
voltage (pd)	V	volt	V
resistance	R	ohm	Ω
capacitance	C	farad	F
energy	E	joule	J
power	P	watt	W
frequency	f	hertz	Hz
wavelength	λ	metre	m
period	t	second	s
temperature	T	degree Celsius	°C
radioactive activity	A	becquerel	Bq
absorbed dose	D	gray	Gy
equivalent dose	H	sievert	Sv
focal length	f	metre	m
lens power	P	dioptre	D
speed, velocity	v	metre per second	m/s
acceleration	a	metre per second squared	m/s^2
force	F	newton	N
weight	W	newton	N
gravitational field strength	g	newton per kilogram	N/kg
work	E_w or W	joule	J
specific heat capacity	c	joule per kilogram per degree Celsius	J/kg°C
specific latent heat	l	joule per kilogram	J/kg
gravitational acceleration	g	metre per second squared	m/s^2

Prefix	Symbol	Value
tera	T	10^{12}
giga	G	10^9
mega	M	10^6
kilo	k	10^3
deci	d	10^{-1}
centi	c	10^{-2}
milli	m	10^{-3}
micro	μ	10^{-6}
nano	n	10^{-9}
pico	p	10^{-12}

Answers

Telecommunications

1. Energy.
2. A = amplitude, B = wavelength.
3. Reflection, refraction and diffraction.
4. B
5. Light travels much faster than sound.
6. a) A b) C c) B d) D
7. Transmitter
8. Almost equal to the speed of light in air.
9. UHF radio waves
10. Vision and sound decoders.
11. Red, green and blue.
12. 24 hours
13. Number of waves increase.
14. Makes a stronger electrical signal.
15. By T.I.R., total internal reflection
16. Aerial.
17. a) 0.5Hz. b) 1.5m/s c) 0.3m
18. Catches more energy waves and reflects these to the receiver at the focus
19. Orbit is such that the satellite is always above the same point of the earth's surface.
20. $f = \dfrac{v}{\lambda} = \dfrac{3 \times 10^8}{0.025}$
    $= 1.2 \times 10^{10}$ Hz = 12GHz
21. Modulation
22. $t = \dfrac{1\,000\,000}{2 \times 10^8} = 0.005s$
23. Different numbers of electrons strike the red and green phosphor dots.
24. Electrical signals in copper wire.
25. The TV signals are high frequency which do not diffract around hills well.

Using electricity

1. They repel
2. They attract
3. Protons, neutrons and electrons; electrons.
4. They can be turned on and off.
5. a) Circuit A is complete, Circuit B is incomplete.
   b) Place the material across the gap in the circuit. If the bulb now glows the material is a conductor, if it does not it is an insulator.
6. The outer casing.
7. Parallel circuits
8. Larger current, more turns in the coil, stronger magnetic field. The split ring changes the direction of the current in the coil every half turn.
9. 3V
10. 6 units, 66p.
11. Direct current is in only one direction; alternating current flows back and forth i.e. it changes direction.
12. $P = VI = 230 \times 5 = 1150$ W.

13. $R = \dfrac{V}{I} = \dfrac{6}{0.25} = 24\Omega$
14. $V = \dfrac{P}{I} = \dfrac{3000}{13} = 230V$
15. $I = \dfrac{P}{V} = \dfrac{1800}{230} = 7.8A$
    so a 13A fuse would be used.
16. 2
17. Fluorescent tube gives greater light as it is more efficient – less heat.
18. Dimmer switch or volume control
19. $E = P \times t = 24 \times (5 \times 60) = 7200J$
20. a) 4 Sidelights: $I = \dfrac{P}{V} = \left(\dfrac{4 \times 5}{12}\right)$
    = 1.7A, 2 Headlights: $I = \dfrac{P}{V}$
    $= \left(\dfrac{2 \times 21}{12}\right) = 3.5A$, Total I = 5.2 A
    b) The bulbs have the same voltage but headlight has higher power, draws more current, so has the lower resistance
21. a) Force is downwards
    b) Force is upwards again
22. a) Smoother
    b) Can use a.c. supply – magnetism reverses as current reverses.
23. a) $P = \dfrac{V^2}{R}$, $R = \dfrac{V^2}{P} = \dfrac{230^2}{60} = 882\Omega$
    b) 441Ω

Health physics

1. Vacuum
2. a) The reflection of a sound wave
   b) 750m
   c) An echo would be heard sooner.
3. It slows down and bends towards the normal; it slows down but does not change direction.
4. Sounds with frequencies so high they cannot be heard by human hearing; prenatal scanning.
5. a) The critical angle.
   b) i) The ray is refracted and partially reflected ii) The ray is totally internally reflected.
6. It always strikes the inner surface at an angle greater than the critical angle and so is always totally internally reflected; an endoscope or cable television.
7. Ear defenders are worn over the ears to prevent damage to hearing. Workers using noisy machinery should use them.
8. a) Beta b) Gamma c) Alpha
9. a) Gamma b) Beta c) Alpha
   d) Gamma
10. Radiotherapy, sterilisation of surgical instruments, radioactive tracers.
11. Alpha radiation cannot penetrate skin.
12. 34–42°C
13. 20–20 000 Hz
14. Ultra-sound
15. Stethoscope

16. a) Rays focus in front of retina.
    b) Diverging lens.
17. Smaller and inverted
18. $A = \dfrac{n}{t} = \dfrac{150}{5 \times 60} = 0.5Bq$
19. Activity decreases.
20. $f = \dfrac{1}{P} = \dfrac{1}{60} = 0.017m$
21. The bulb is not a distant object and the rays will not be parallel.
22. a) Opposite side of the body to the transmitter
    b) Same side as the transmitter.
23. Distance, shielding, handling tool.
24. 1200 → 600 → 300 → 150 → 75 Bq
25. Energy and type of radiation.

Electronics

1. Input, Process, Output.
2. Electrons.
3. Microphone, thermistor, LDR.
4. Loudspeaker, LCD display, LED.
5. This is a light emitting diode.
6.

7. Analogue and Digital.
8. Gain = $\dfrac{\text{output voltage}}{\text{input voltage}} = \dfrac{1.8}{0.02} = 90$
9. Analogue.
10. Digital.
11. LED is the wrong way round.
12. A resistor.
13. OR gate.
14. AND gate.
15.
16. 0 0|0
    0 1|0
    1 0|0
    1 1|1
17. A resistor whose resistance decreases in bright light; controlling street lights.
18. LDR, resistance decreases when illuminated.
19. An electronic switch.
20. When the voltage to the base of the transistor increases to above a certain value (about 0.7V) the transistor is switched on and conducts.
21. $R = \dfrac{V}{I} = \dfrac{5 - 2}{0.015} = 200\Omega$
22. Capacitor, Inverter and Resistor.
23. Decrease the value of the capacitor or resistor.
24. Binary.
25. It is made from 7 LEDs or LCDs.

Answers

Transport

1. 10m/s
2. 270km
3. 25s
4. With tiredness the stopping distance will increase.
5. 50J
6. Slow down, speed up and change direction.
7. It is streamlined so there is less water resistance.
8. Air resistance will equal weight; the sky diver will fall at his terminal velocity.
9. 5m/s^2
10. a) X = reaction, Y = weight, Z = friction
    b) All the forces acting on the object are balanced.
    c) 30N
    d) Add a lubricant (oil or water); put the object on wheels.
11. a) 50m b) 800N c) 40 000J
    d) 1600W.
12. a) Both land together.
    b) a = 10m/s^2
13. $E_p = mgh = 80 \times 10 \times 500 = 400\,000$J
14. a) $\frac{60}{10} = 6$ m/s^2
    b) Resistive forces balance engine force.
15. a) $a = \frac{12}{3} = 4$ m/s^2
    b) $F = ma = 50 \times 4 = 200$N
16. a) $E_W = F \times d = 400 \times 20 = 8000$J
    b) 200W
17. a) $a = \frac{v - u}{t} = \frac{8 - 0}{2} = 4$m/s^2
    b) $d = (\frac{1}{2} \times 8 \times 2) + (8 \times 20)$
       $= 8 + 160 = 168$m
    c) pushing force greater at start, balanced forces during steady speed.
    d) crouch and wear smooth clothing
18. a) $E_p = mgh = 50 \times 10 \times 1.2 = 600$J
    b) $P = \frac{E}{t} = \frac{600}{8} = 75$W
19. a) $\Delta v = at = 4 \times 3 = 12$,
       $v = 35 - 12 = 23$m/s
    b) $F = ma = 80 \times 4 = 320$N.
20. E_K before $= \frac{1}{2}mv^2$
    $= \frac{1}{2} \times 1000 \times 10^2 = 50\,000$J,
    E_k after $= \frac{1}{2}mv^2 = \frac{1}{2} \times 1000 \times 20^2$
    $= 200\,000$J,
    so the difference $= 150\,000$J.
21. E_P lost $= E_K$ gained, $mgh = \frac{1}{2}mv^2$,
    $10 \times 6 = \frac{1}{2}v^2$,
    $v = \sqrt{120} = 10.95$m/s

Energy matters

1. a) Electric motor
   b) Generator or dynamo
   c) Solar cell
2. Gravitational potential energy
3. a) Radiation b) Convection
   c) Conduction
4. a) An energy source which will not be exhausted.
   b) Wind, wave, tidal, solar.
   c) coal, oil, gas.
5. Use renewable sources of energy; insulate homes and factories to reduce energy wastage; develop more efficient machines and generators.
6. Loft insulation, cavity wall insulation, double glazing, underlay and carpets, draft excluders and curtains.
7. 60%
8. 40J
9. a) Gravitational potential energy
   b) Kinetic energy
   c) Water is stored in the top loch until electrical energy is needed.
   d) 75%
10. Roof, 25%
11. 250J
12. 50 turns
13. Transformers require a.c. for a changing magnetic field, batteries give d.c.
14. No change
15. E = cmΔT = 1000 × 400 × 15 = 6 000 000J = 6 MJ.
16. To distribute electrical power around the country by transmission lines.
17. Heat is a form of energy, temperature is a measure of hotness.
18. a) $t = \frac{E}{P} = \frac{cm\Delta T}{P}$
       $= \frac{4180 \times 0.25 \times 15}{50} = \frac{15\,675}{50}$
       $= 313.5$ s
    b) The cotton wool is an insulator and reduces heat loss by conduction, the metal foil is a good reflector and reduces heat loss by radiation.
19. $E = ml = 0.1 \times 3.34 \times 10^5$
    $= 3.34 \times 10^4$J
20. a) $P_{in} = P_{out}$, so $V_P I_P = V_S I_S$.
       $230 I_P = 115 \times 16$,
       so $I_P = \frac{115 \times 16}{230} = 8$A.
    b) Efficiency $= \frac{\text{Output Power}}{\text{Input Power}} \times 100\%$
       $= \frac{115 \times 16}{230 \times 12} \times 100 = 66.7\%$.
21. Heat is generated in the coils and the core of the transformer.

Space physics

1. Gravitational forces
2. An object that orbits a planet.
   a) The Moon b) A weather satellite
3. Violet, spectrum.
4. 8 minutes.
5. 100 000 years
6. Sun, moons, planets.
7. The universe.
8. An object which orbits a sun.
9. Violet.
10. Red.
11. Objective: produces an image of the object being viewed. Eyepiece: magnifies the image produced by the objective lens.
12. Radio wave / x-rays
13. The elements which are present in the star.
14. Spectroscope.
15. Thrust and Weight.
16. Thrust is greater than weight, giving an unbalanced force upwards.
17. $W = mg = 160 \times 1.6 = 256$N.
18. The distance travelled by light in a year.
19. The greater the diameter, the more light is captured and the brighter the image.
20. Gamma rays, X-rays, ultra-violet, visible light, infra-red, microwaves, TV and radio.
21. If A exerts a force on B, B exerts an equal and opposite force on A.
22. Frictional forces cause the craft to heat up.

Index